Electric Galaxies and Solar Systems

by Rolf A. F. Witzsche

Contents

About the Illustrated Science series
On the Ice Age and Climate Change
and the book

Electric Galaxies and Solar Systems

Mainstream cosmology regards the universe, the galaxies, and the solar system exclusively organized by gravitational force that is known to be the weakest universal force. The next higher-order force in the universe is the electric force that is 39 orders of magnitude stronger than the gravitational force. However, it is not allowed to be recognized as an organizing force in the universe, because this force is expressed in electrically charged plasma that is deemed not to exist in mainstream cosmology, which thereby imprisons itself with cosmo-mythologies where nothing is actually true.

While technology has furnished astronomy with amazing capacities for looking at the universe, ironically, what is observed is being falsely interpreted on the basis of assumptions that are simply not true, that are mythological assumptions. As a consequence, ironically, mainstream astronomy looks at the universe blindfolded. What comes out of it, of course, are tragic misperceptions. The results are often so confusing that mysterious fudge factors need to be invented to make the results appear plausible. No such fudge factors are needed in plasma cosmology.

With the next Ice Age on the near horizon, potentially beginning in the 2050s, we cannot afford to play games with fudge factors. The recognition of the true nature of the universe, the galactic system, and the solar system, that together drives the Ice Age dynamics, becomes an existentially critical issue. If humanity remains 'asleep' on this front, we may all die in the easy chair of the consequence when the glaciation conditions resume, which evidence promises, will happen quickly.

Plasma in the physical universe is as challenging in perception as the spiritual domain in the human sphere. Both are invisible, except by their effects, but they are understandable and knowable. But how does one break away from the fairy tales that inspire delusions? Answers must be found.

With the Ice Age Challenge now before us, we face two imperatives. One is to understand the real physical dynamics that power and affect the Sun, and with it to create the physical infrastructures that enable human living to continue in an Ice Age climate. The second challenge, and this is the greater challenge, is to raise up our humanity to such height as will impel us to get the job done. Some say that miracles are needed on both fronts. But what of it? Are we, as human beings, not the miracle makers on the Earth?

In the real universe, the cosmic operations are anti-entropic in nature, and expanding and progressing. We, ourselves are evidence of this progression. Should this progression have ended? Neither is our Sun isolated from the progressive nature of the universe, but expresses its dynamics, its resonating plasma streams, and their reflection in the climate on Earth. Shouldn't we develop ourselves spiritually and culturally, likewise?

Climate Change reflects the nature of the universe. It should also be reflected in us.

The Earth itself is the creation of the Sun, with its atoms having been massively synthesized in high-energy times near the center of the galaxy.

The synthesizing plasma fusion is presently at a low state, though it is currently enhanced for our Sun by electromagnetic 'Primer Fields' that focus interstellar plasma onto the Sun in a highly condensed manner. When the plasma-focusing system becomes inactive, below the required threshold conditions, the Sun reverts to a type of cosmic default level with 70% less energy being radiated, and higher rates of solar cosmic-ray flux being experienced.

At the present rate of plasma diminishment being experienced, the solar activity phase-shift threshold to the next Ice Age period may be crossed in 30 years, or in the 2050s, most likely. With the primer-fields system gone inactive by then, the climate on Earth will get 40 times colder than the Little Ice Age in the 1600s had been. Ice core evidence promises that.

Without the needed preparations for human living in such an environment, 99% of humanity would die of starvation, both by the cold, and by CO2 depletion that diminishes agriculture, as more CO2 becomes dissolved into the sea.

With the 'Primer Fields' being critical for our very existence, the exploration of them is likewise critical.

In the Little Ice Age, between 10% and up to 30% of the populations in Europe had perished by starvation. The last Big Ice Age was evidently vastly harsher. Only 1-10 million people emerged from it alive. That's all we had after 2 million years of development. We want to do far better this time around; and we can, with large-scale technological infrastructures for our food supply. But will we create them? Will we get the job done in the 30 years that we still have left before the Ice Age starts anew? Will we even consider it? And how certain are we that the phase shift to the next glaciation period will begin, as the evidence suggests, in the 2050s? We have no slack on this front. Should we fail us on this absolute front, we would be committing suicide.

Numerous fields of evidence tell us that the next Ice Age is near. That's where the truth begins. Most of the evidence was discovered in the 1990s and thereafter. Some evidence is measured in ice cores; some is measured in space, by satellites. Some measurements are also made on the ground in terms of measurements of the Earth's magnetic-pole drift observed in northern Canada. All of this is seen combined with high-energy physics experiments at a leading national laboratory, and is also explored in the small in static experiments.

So, what will the answer be? Will we move with the evidence? Or will we lay ourselves down to die by default?

It takes an independent researcher to brake the taboos that have kept mainstream cosmology imprisoned, increasingly, during the past century, even while what is regarded as taboo is known to be wrong.

The Illustrated Science series is intended to open the scene beyond the threshold of accepted taboos, to where the actual physical evidence speaks for itself.

The scope of the existential challenge that the Ice Age brings with it, takes astrophysics out of the academic domain and places it into the foreground as one of the most-critical issues of our time. The big Climate Change events that have already worldwide effects are mere fringe effects in the flow of the ever-changing cosmic dynamics. The big effect, when the Ice Age begins anew, promises to be caused by a dimmer and colder Sun. The loss of 70% of the Sun's radiated energy defines our climate future that begins in the near term.

Sure, we can live with all that by creating new platforms for agriculture that are able to operate under Ice Age conditions. But will we do it? The task is enormous. Or will we fail ourselves on this front? We have no reason to allow us to fail. We have the materials and energy resources on hand to accomplish everything that is required for us to continue to live in an Ice Age World. But will we do it? The big question that never goes away, therefore, is; will we develop our inner resources as human beings sufficiently to get the job done, and to get it done in time? Or will we do nothing, ignore the challenge, and condemn our children and one-another to an agonizing death by starvation? That's the choice.

Towards meeting the inner challenge, I have created the epic series of novels, The Lodging for the Rose. And further, towards meeting the science challenge, I have produced numerous research books and several dozen exploration videos that the Illustrated Science series is modeled after. The work is the result of a quarter century of research, for which numerous elements of evidence in related fields came to light during the timeframe of my research.

It is my hope that the work that went into all of these projects will help in some degree - for humanity that we are all a part of - to write itself a ticket to have a future.

High-resolution color images, of the images in this book, can be obtained at www.iceagetheatre.ca

Expressions of a common principle

** One Principle, One Universe, Many Expressions
It would be surprising if Universal Principle was not expressed
universally in widely varied forms, and without exceptions.
In the real universe, no 'surprise' is found. All evidence points to
one single basic platform that is expressed in the forming and
operation of Galaxies in the larger context, and Solar Systems in the
smaller context, and in planetary systems in the very small context,
with no fundamental differences between them, even in the many
forms of expressed characteristic, such as one finds in differences in
scale. In the fundamental sense, the universe and its multi-level
physical expression is built on the platform of a single, common,
universal platform. It seems hard to accept that this should be so in
the mental landscape where the opposite is being taught.
In the context of science, we face two different worlds as it were.
One of these is the Old World that has been promoted as absolute
truth from the 1900s on, which is deemed a practical world were
gravity is king and a falling apple can hit one on the head. It is being
said in this Old World that everything that can be discovered has

been discovered. In this context it can be said that each of the four phenomena shown here are as isolated in function from each other, and different in design, as night and day, except for the one doctrine that rules them all: the master doctrine in which gravity is the king of the Universe.

But science has advanced, hasn't it?

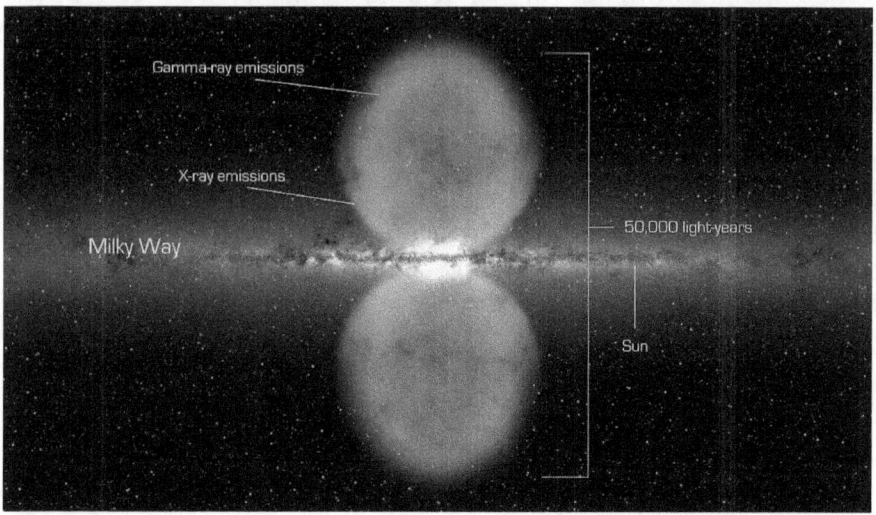

But science has advanced, hasn't it? With advanced technology in instrumentation, some surprises have popped into view that violate everything that is held dear in the box of the Old World doctrine. NASA has discovered, by looking at the world in x-ray and gamma-ray light, that two gigantic plasma bulbs extend above and below our Milky Way Galaxy, perpendicular to the galactic disc, for 25,000 light years into opposite directions. While some exotic efforts have been made to attribute the discovered phenomenon to his majesty King Gravity, the gigantic geometry of the plasma bulbs that have been photographed by their extremely high-energy signature in x-rays and gamma-rays, makes it fairly obvious to an open mind that we are dealing with something totally different than a gravity-forced phenomenon.

With this paradox the Old World breaks down, and a New World opens up that puts the entire cosmos into a new light. While crossing the threshold into the New World the realization begins to dawn that, maybe, we don't know everything yet, so that it is still possible to make new fundamental discoveries, such as the recognition that the cosmos is not empty space, but is pervaded with streams of plasma that the plasma structures shown here are an affect of, which by this single profound recognition revolutionizes our perception of how a galactic system actually operates, and therefore, how the universe itself operates.

Path into the New World

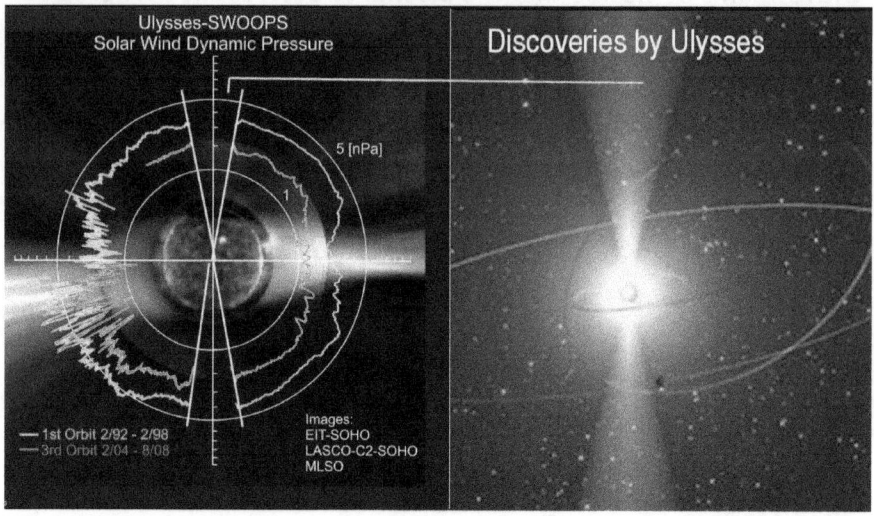

A measure of progress on this path into the New World gradually opens up before us. We know that progress has been made when the Old World becomes conspicuously absent in the face of evermore evidence with which the old doctrines become irrelevant. With advanced technologies we have begun to discover evidence that shouldn't exist, according to what is believed in the Old World.

It is here, as we boldly step from the Old into the New World and explore its vistas, that grand stories about ourselves come to light in which we discover a bit more of ourselves as human beings. In discovering new frontiers, and in coming to understand bit by bit some of its amazing aspects that had remained unseen before, a new face of man comes to the foreground that in the Old World has been kept shrouded for far too long, which had thereby remained unrecognized.

The Cassini-Huygens spacecraft

Saturn eclipsing the Sun

In the face of science, we discover elements of our humanity as human beings that makes one proud to be human, even while the discoveries don't seem to have any direct practical relevance. We become proud of ourselves for our extreme capability to see what no other form of life has ever seen, such as Jupiter eclipsing the Sun. Some say that such technological ventures are financially not practical. I would say that these ventures are extremely efficient in raising the cultural optimism in society, even to the point of raising the status of man as the new measure for life itself that takes us far beyond the measures of practicality.

And the most amazing part of the picture that you see here, is that it was seen in the human mind before it was seen with the eyes of a camera as a means of the technological extension of our eyes. The Cassini-Huygens spacecraft that has photographed the planet Saturn eclipsing the Sun, began to be planned as far back as the 1980s. The completed spacecraft was eventually launched 17 years later, in 1997, towards a specific point in space where the planet Saturn would arrive in 2004 at the same time that the space craft would arrive, based on the scientific discoveries of universal principles by Johannes Kepler back in the early 1600s.

The robotic spacecraft at Saturn

The robotic spacecraft was designed to separate itself on arrival at Saturn, and land a part of it on Saturn's moon Titan, to both explore the moon, and for the moon to serve as a base for a relay station for radio transmissions to Earth. The mission planners knew exactly what to expect for the spacecraft and how to position the spacecraft to photograph Saturn with the Sun standing behind it in order that the space around it, which is normally hidden by the solar glare, becomes visible.

Discovery of two extra rings around Saturn

Saturn eclipsing the Sun

The result of the venture was the discovery of two extra rings around Saturn that are made up of microscopic frozen water vapor that is ejected by Saturn's moon Enceladus from a hundred geysers operating in the extreme cold of nearly 200 degrees below zero.

An electric connection exists

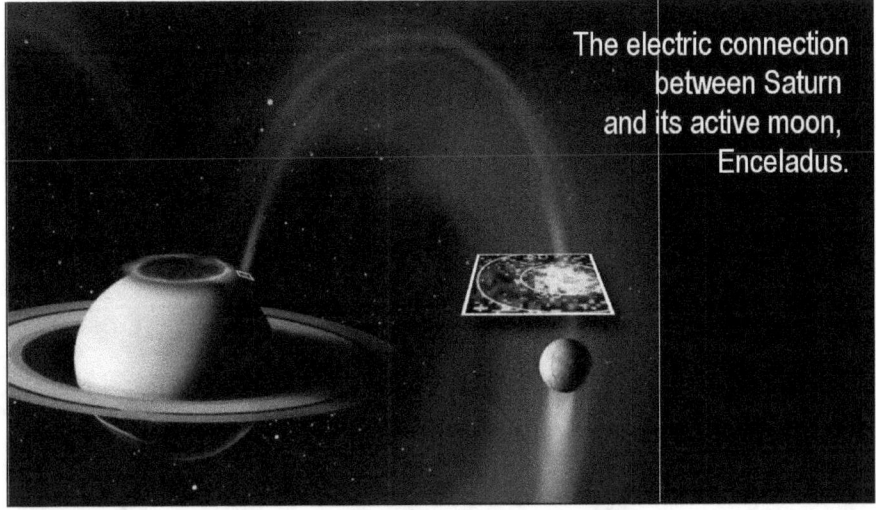

The electric connection between Saturn and its active moon, Enceladus.

It was further discovered by the flying laboratory that the Cassini satellite still is, which is still operating, that an electric connection exists between Saturn and its active moon, Enceladus. The discovered connection originates near the ring of Saturn's own electric connection with the solar system.

It seems that the more we open our eyes to the universe, the more we are confronted with evidence that we live in an electricity-rich universe that we see evidence of wherever we look, from the small to the gigantic.

This artist's concept of the discovered, active, electric moon, shows a glowing patch of ultraviolet light extending from near Saturn's north pole along the path of magnetic connection, to Saturn's moon Enceladus. The footprint and plasma flow field lines are not visible to the naked eye, but they were detected by the ultraviolet imaging spectrograph and other instruments on NASA's Cassini spacecraft. The electric footprint and connection to the moon, discovered by Cassini, presents the presence of an electric plasma circuit that appears to power the geysers erupting on Enceladus near its South

Pole that feed the two faint outer rings?
But can we recognize fully, from the observed phenomena alone,
what principles stand behind the phenomena that we see? No we
don't?

We can recognize that the dynamics of the gigantic structures

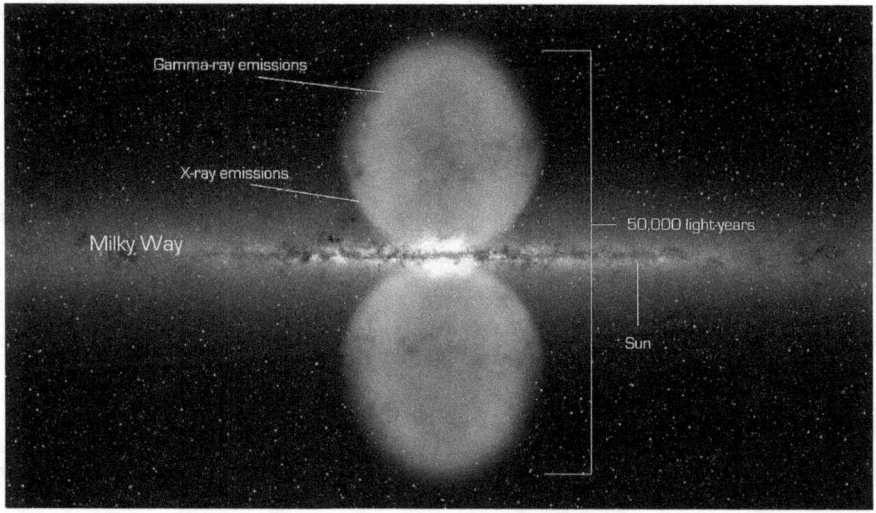

However, we can recognize that the dynamics of the gigantic structures on the galactic scale, cannot be the result of the weakest force in the universe, which is gravity, but is evidently the result of the strongest force of the universe, which is the electromagnetic force that is 39 orders of magnitude stronger with effects that have potentially an infinite range.

In our tiny solar system

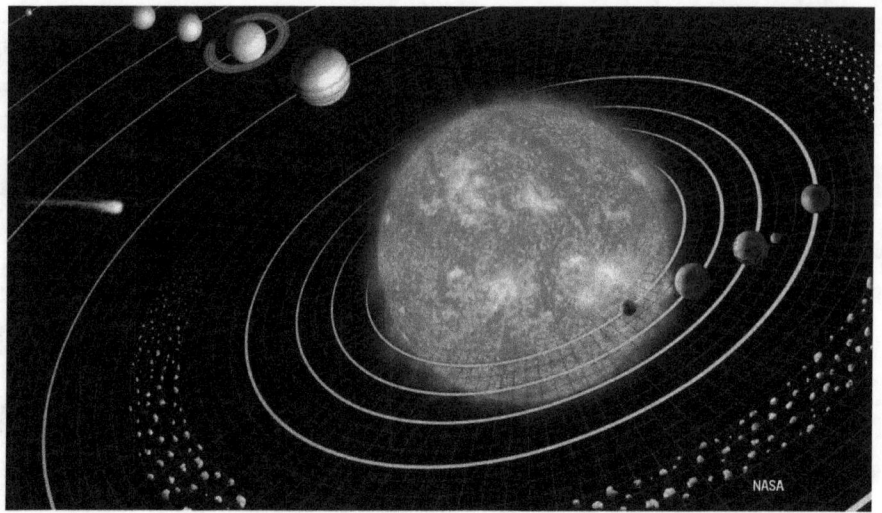

In our tiny solar system, the plasma-flow structure is far-less dense, and therefore completely invisible, while the Sun's gravity is large enough that it has a huge effect on the movements in the solar system. However, the electric plasma effect that stands behind the entire system, can nevertheless be recognized visually by its effect on the solar wind.

*The Ulysses polar orbit around the Sun

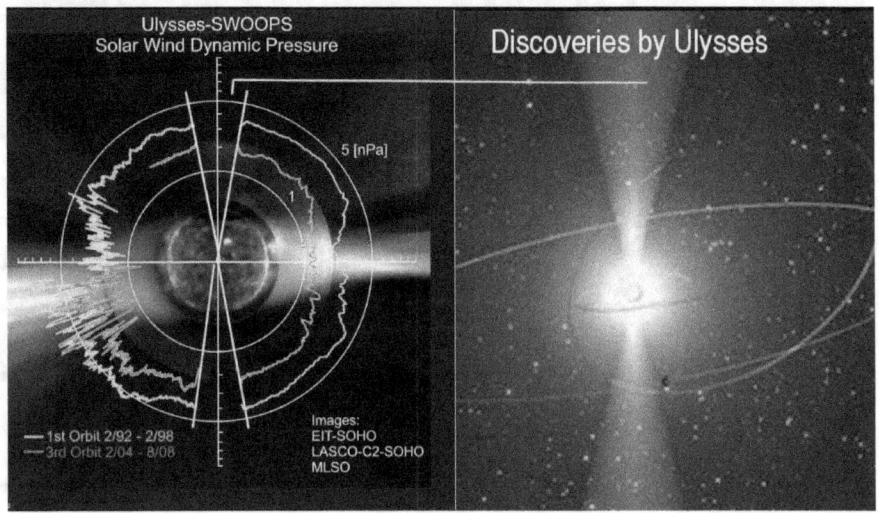

The Ulysses satellite that flew a polar orbit around the Sun has reported a sharply defined void in the solar winds in a small region over the poles. This region corresponds with the immensely concentrated plasma flow structures that have been discovered in laboratory plasma-flow experiments between an electric inflow and outflow. The bowl-type magnetic structure that the flow of electricity had created in the experiment, which were made visible in the experiment, is also visible in space by corresponding effects.

The specific shape of the common geometry

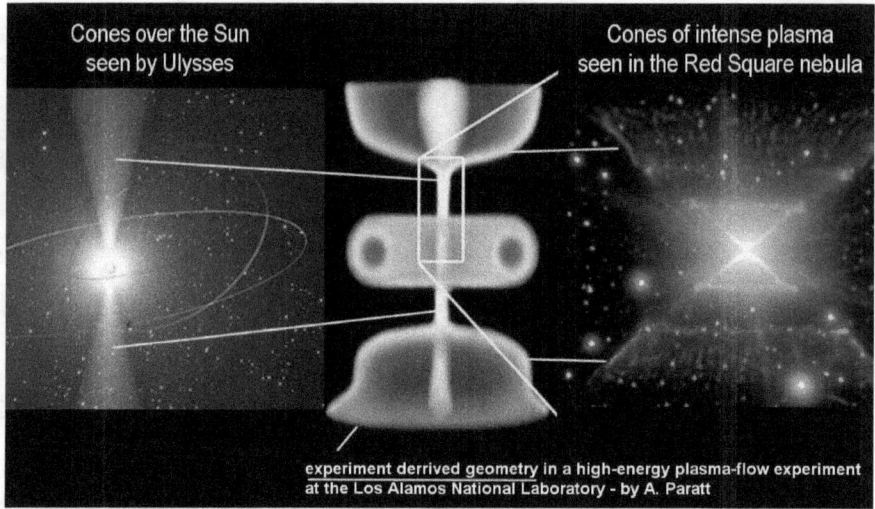

Cones over the Sun
seen by Ulysses

Cones of intense plasma
seen in the Red Square nebula

experiment derrived geometry in a high-energy plasma-flow experiment
at the Los Alamos National Laboratory - by A. Paratt

The specific shape of the common geometry that we see in some
form reflected in both the galactic system and in the solar system, is
evidently in both cases a function of the density of the flowing
electric plasma in their respective system and the subsequent
magnetic fields that the movement of electricity creates within
these systems. All magnetic effects in the universe are the result of
electricity in motion, exclusively.

David LaPoint, who conducted the experiment

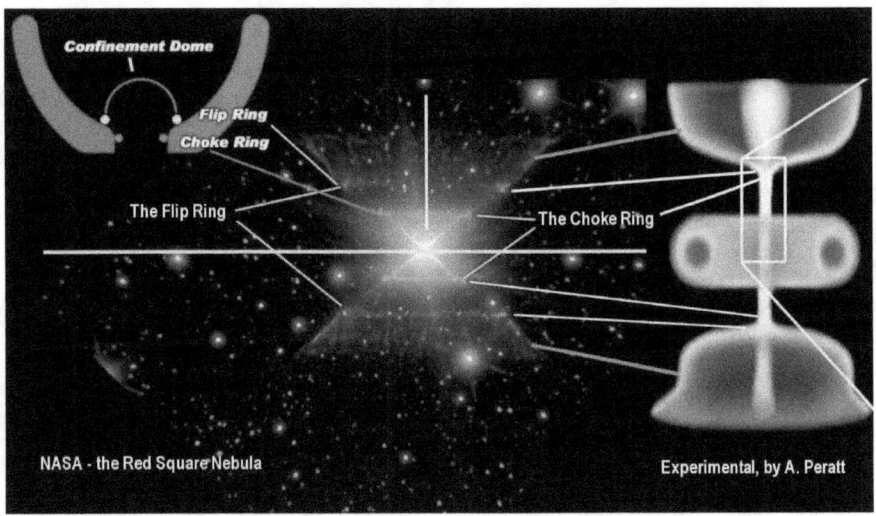

The details of the flow dynamics of the magnetically created structure have also been explored in static experiments. In the highest-intensity region of the magnetic pinch effect, a magnetic ring has been detected that causes inflowing plasma to be flipped backwards, whereby the inflowing plasma becomes magnetically concentrated. David LaPoint, who conducted the experiment, termed the magnetic ring the flip ring, and the magnetic result, the confinement dome.

Plasma structures 'photographed' by NASA

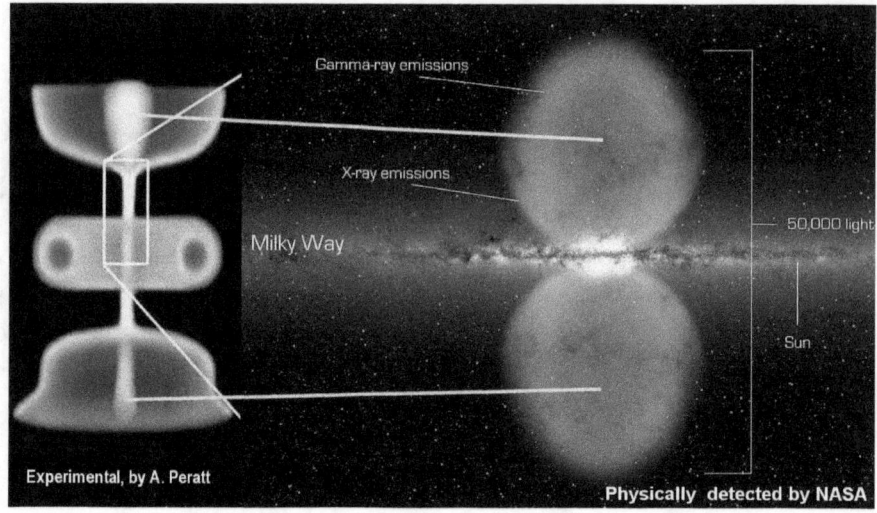

The plasma structures that have been 'photographed' by NASA are very-large-scale examples of plasma being magnetically concentrated under a confinement dome. Below the flip ring, David LaPoint recognized still another magnetic ring structure that pinches the outflow from the confinement dome into a narrowly focused stream.

In the case of our solar system

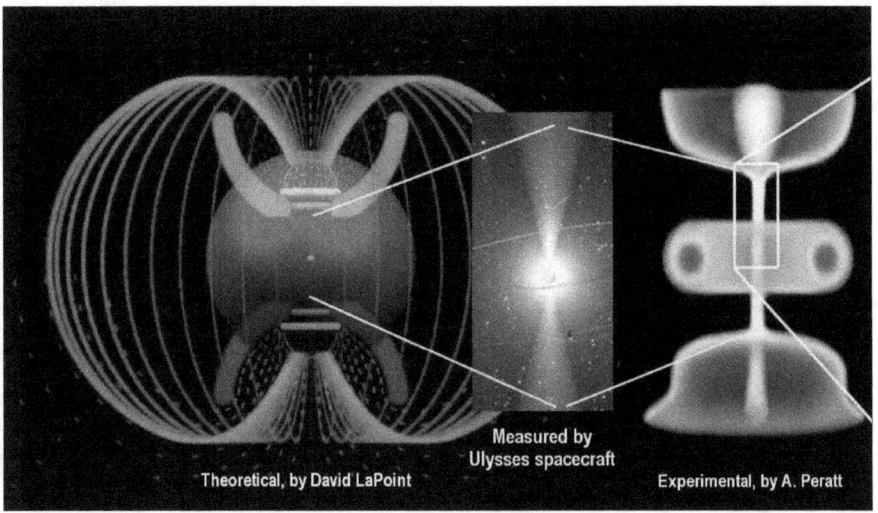

Measured by
Ulysses spacecraft

Theoretical, by David LaPoint

Experimental, by A. Peratt

In the case of our solar system, the plasma stream would be
focused onto our Sun that consumes some of the plasma
surrounding it, in a process of plasma fusion on its 'surface' where
all known natural atomic elements are synthesized, and have been
for as long as the Sun existed. Whatever portion of the plasma
stream that flows through a solar system is not consumed in
nuclear fusion, flows on and out of the solar system in the reverse
of the in-flowing process, that results as an expanding process.

The opposite confinement dome

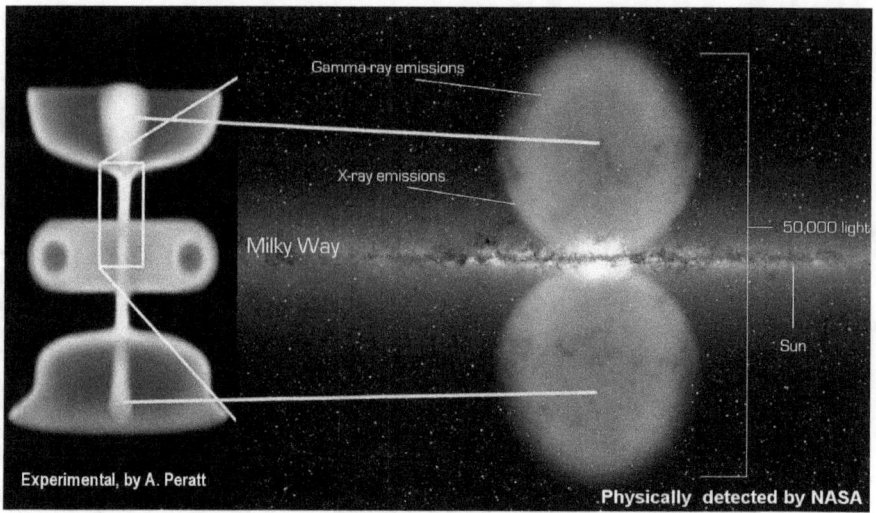

The opposite confinement dome for the galactic system, buffers the outflow and regulates the expansion of the transit plasma into intergalactic space where the plasma stream flows to the next galaxy

The void that Ulysses saw

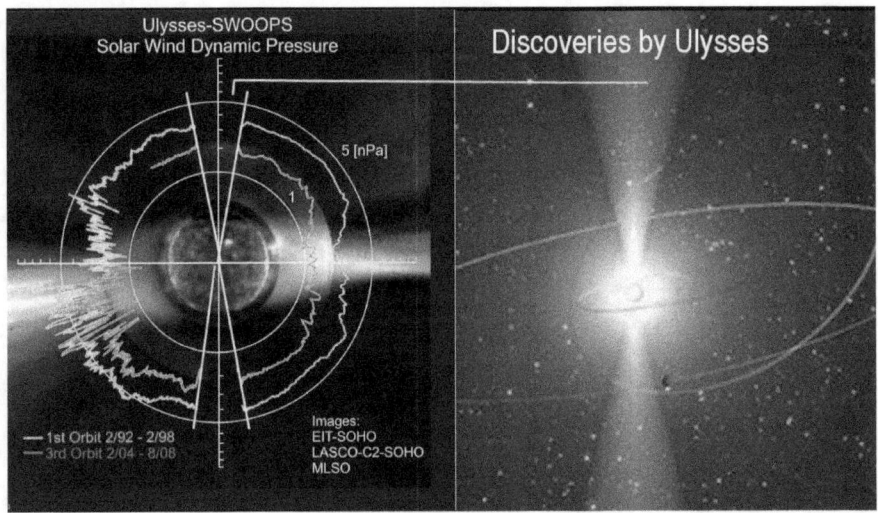

The Ulysses satellite encountered the plasma stream as a conical void in the solar winds, at a distance over the Sun's poles slightly less than the orbit of Jupiter. the voids would be located between the Primer Fields as David LaPoint has named the magnetic feature formed by self-pinching plasma streams. The void that Ulysses saw stands as proof that we live under an electric sun in an electric universe.

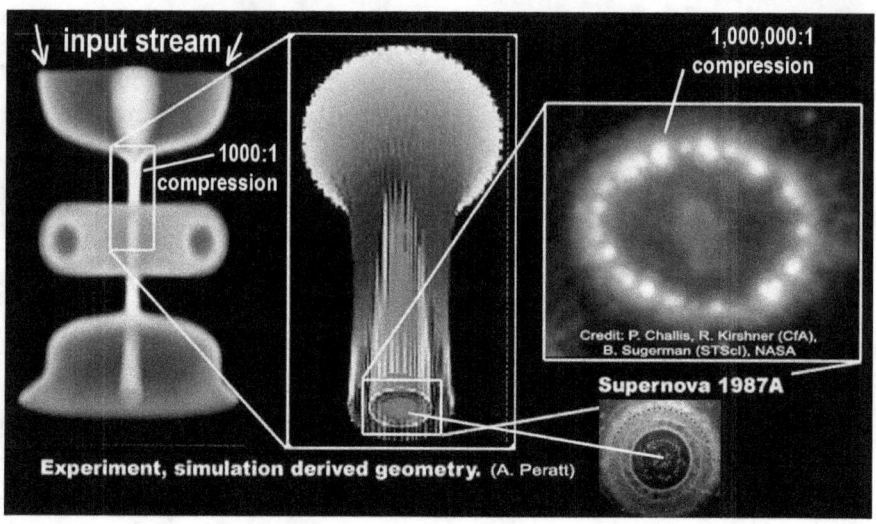

input stream

1000:1 compression

1,000,000:1 compression

Credit: P. Challis, R. Kirshner (CfA), B. Sugerman (STScI), NASA

Supernova 1987A

Experiment, simulation derived geometry. (A. Peratt)

** Do We Really Live in an Electric Universe?

This big question, and similar questions, have been explored with big experiments in the laboratory, such as the high-energy plasma-flow experiments that were conducted at the Los Alamos National Laboratory under the direction of Anthony Peratt. Normally, electricity is invisible. The two types of plasma particles that the universe is made of, that carry an electric charge, are far too small to be seen directly. Nor are plasma particles by themselves capable of emitting light. But they do become visible by their effect on atoms in their path. The atoms become energized by the electric interaction with plasma. As a result, atoms do emit light according to the intensity or density, of the plasma flowing around them. Some researchers at the Los Alamos National Laboratory, as the result of experiments, have come to the recognition that 99.999% of the mass of the universe exists in the form of plasma that is invisible, which becomes visible only when the plasma concentration is extremely dense and strongly interacts with atomic elements that reveal the movement of plasma by emitting light.

The principle discovered in laboratory experiments

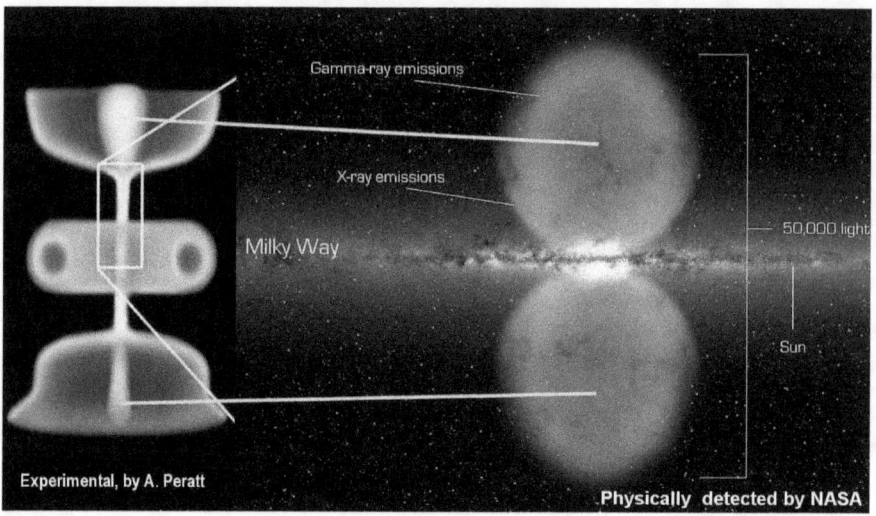

The principle discovered in laboratory experiments enable us now to recognize in the mind what it is that we look at, when our high-technology imaging capability reveals gigantic plasma structures over our galaxy. In the old box of constricted science where electric plasma is not recognized to even exist.

Galaxy M82   Milky Way Galaxy

The plasma domes are explained as jets of energy being emitted from super-massive black holes at the center of the galaxy for which no causative principles are known to exist, nor actual evidence for them that accords with scientifically known facts. In contrast, it is far more creatively efficient to correlated the visible phenomena with experimentally discovered physical facts.

A number of science experiments

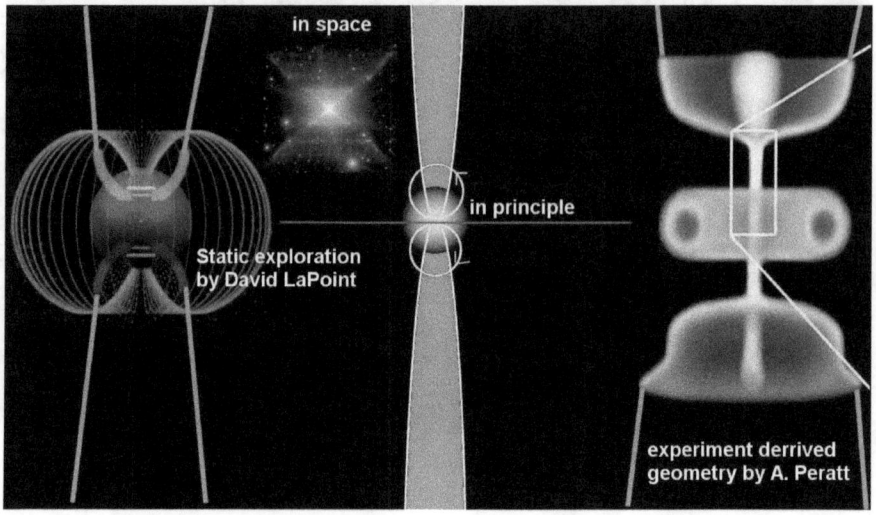

A number of science experiments with different types of laboratory experimentations have shed new light on what it is that we see in space in terms of operating principles. The discoveries, of course, enable more truthful perceptions of the operation of the solar system that our Earth is a part of, and this all based on hard evidence derived from physical experimentation and measurements.

Our spacecraft experiments

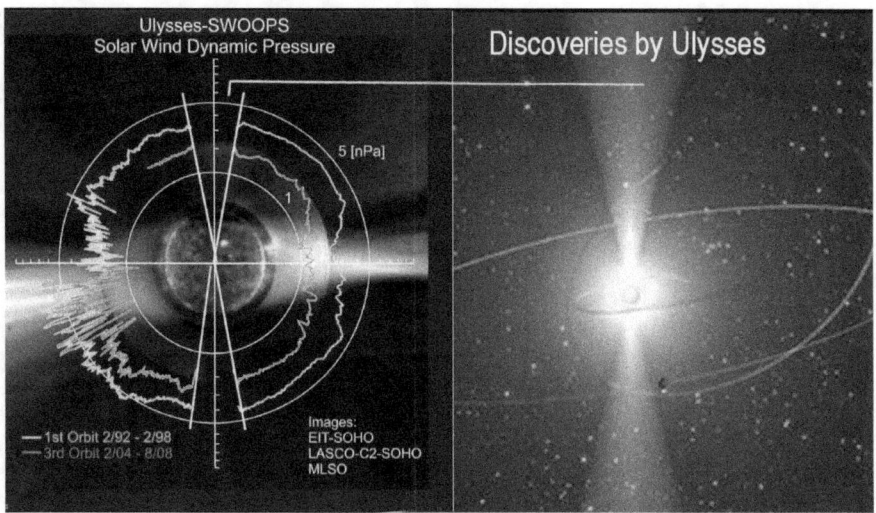

Our spacecraft experiments have measured facts high above the polar regions of the Sun that no eye has ever seen, but which the mind can subsequently relate to, based on laboratory observed facts.

When one combines the spacecraft's measured effects

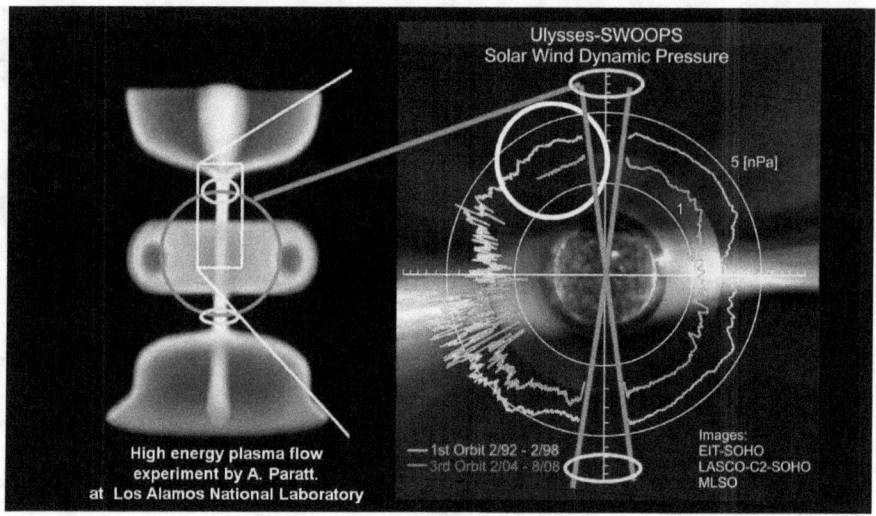

High energy plasma flow
experiment by A. Paratt.
at Los Alamos National Laboratory

Ulysses-SWOOPS
Solar Wind Dynamic Pressure

5 [nPa]

— 1st Orbit 2/92 - 2/98
— 3rd Orbit 2/04 - 8/08

Images:
EIT-SOHO
LASCO-C2-SOHO
MLSO

When one combines the spacecraft's measured effects, where the solar wind is interrupted precisely in the regions where laboratory experiments tell us where we should expect to see the plasma connection of the Sun with interstellar plasma flow, then the multifaceted physical evidence speaks to us of a Sun that is entirely electrically powered by interstellar plasma streams flowing into the Sun. The evidence stands before us like a New World opening up. We can ignore the evidence, of course, and keep our eyes closed, but in doing so we cheat ourselves as we thereby close the door to the recognition of the effects that the electric Sun produces on the climate on Earth, with the occurrence of ice ages, and so on.

By being truthful with ourselves

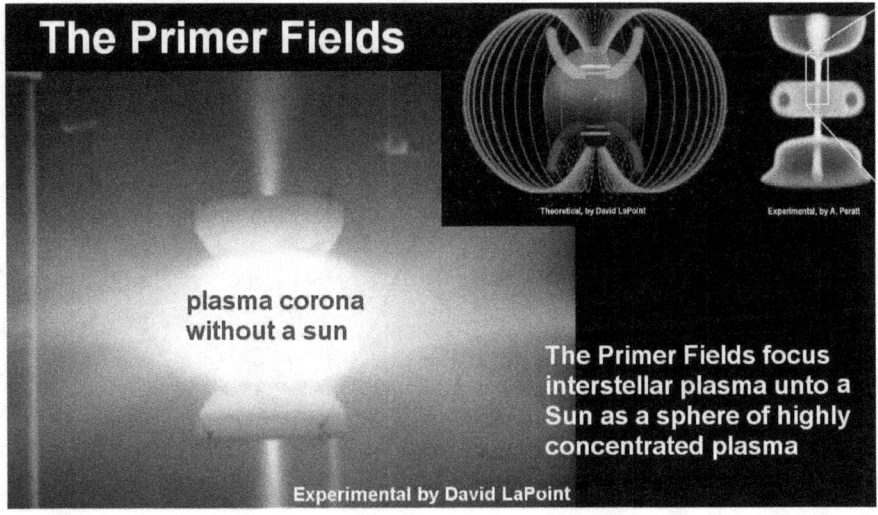

The Primer Fields

Theoretical, by David LaPoint

Experimental, by A. Peratt

plasma corona
without a sun

The Primer Fields focus
interstellar plasma unto a
Sun as a sphere of highly
concentrated plasma

Experimental by David LaPoint

Moreover, by being truthful with ourselves, in exploring the
dynamics of the phenomena that we see, the entire solar system
comes to light in our mind with a more truthful perception of it. If
we fail ourselves at the open door that advancing discoveries set
before us, we literally deny ourselves as human beings, as an
expression of the Universe itself with the built-in freedom to step
out of the box of old constricted perceptions to evermore
discoveries of the unseen. This is what David LaPoint has done by
exploring with essentially 'static' experiments the electric principles
that are expressed in the magnetic primer fields in the universe. His
work reflects in principle the results of the dynamic high-power
experiments produced in the lab by Anthony Peratt, as one would
expect, because the principle of the Universe is singular and is
expressed universally in all forms and in all systems. Consequently,
all forms operate in the universe as a single, interacting, dynamic
unit.

When the evidence is ignored, of the electric dynamics

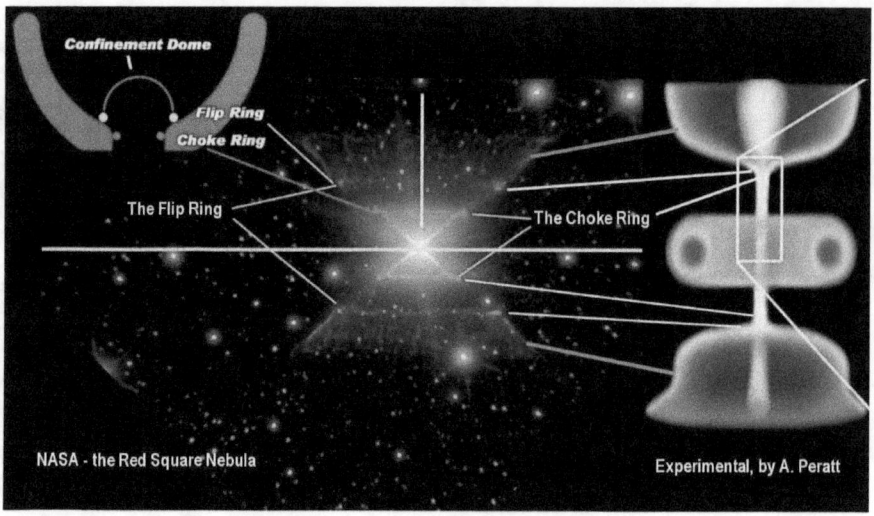

The danger is that when the evidence is ignored, of the electric dynamics that operate at all levels of the universe, the consequences are also being ignored that correspond with leading-edge research in cosmic electro-dynamics. The resulting loss is tragic, because the recognition of the dynamic results is critical to the very survival of most of humanity existing on Earth, for the simple reason that in the electric dynamics of the universe, the ice age dynamics are rooted.

Next Ice Age to occur in 2050s

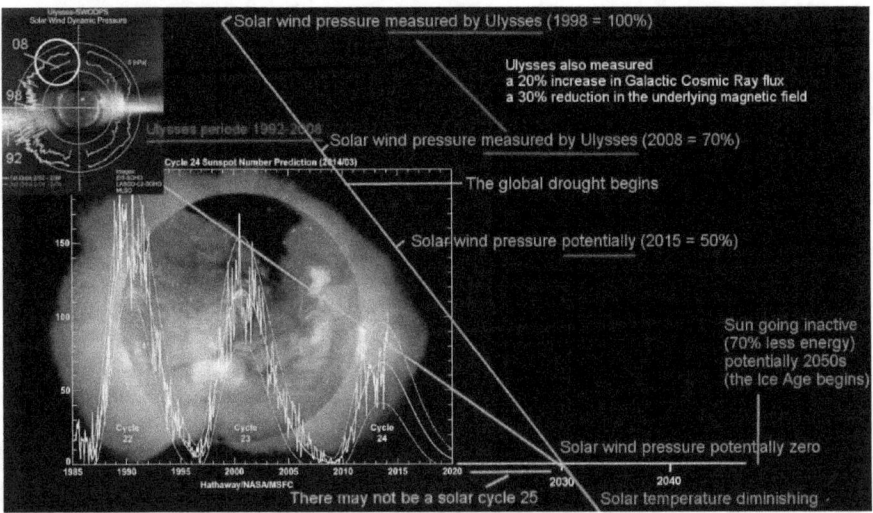

The evidence that has been measured so far with our instrumentation and our exploration of the principles involved, project the start of the Next Ice Age to occur in 2050s timeframe with a 70% loss of solar energy being radiated. That's the evidence speaking.

The consequence of ignoring the critical evidence would result in the loss of most of the agriculture in the world today, when the areas outside the tropics become uninhabitable as the Next Ice Age begins with an extremely short transition in the process. Inversely, if we were to respond to the evidence we have already at hand, that we see reflected everywhere, we would create vast new infrastructures for our continued living in the tropics.

We would create 6000 new cities across the tropics

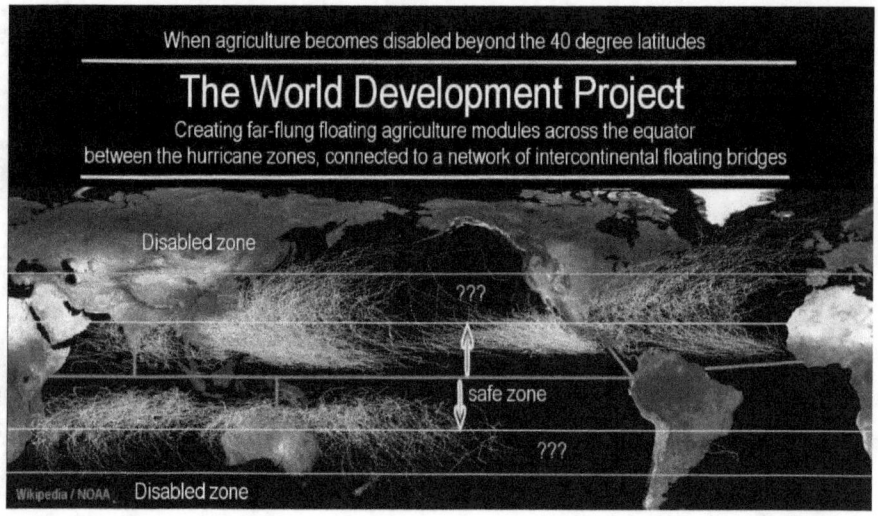

When agriculture becomes disabled beyond the 40 degree latitudes

The World Development Project

Creating far-flung floating agriculture modules across the equator between the hurricane zones, connected to a network of intercontinental floating bridges

We would create 6000 new cities across the tropics, with new industries and new agriculture, most of it laid out afloat onto the tropical seas. The evidence that is presently unfolding for the future should prompt such an intensive worldwide creative response, that we would guarantee with it that we will have a future. The evidence that prompts this, is not small by any means, and is widely apparent in the operation of the galactic systems, solar systems, planetary system, and even the climate systems that affect our living on the tiny planet called Earth. The power for living that we are able to gain with our remarkable intellect as the gem of the Universe with the sparkle of a diamond, brilliant in in life, culture, creativity, and productivity, renders us the absolute pinnacle of life on the planet with the capacity to uplift and protect even the biosphere on Earth during the harsh times ahead when the Ice Age begins anew. That's when our sparkle as a diamond will make the difference for life on Earth between living abundantly, and mere survival on a small and primitive scale near the precipice to extinction.

Let's explore some of the details of the electric evidence

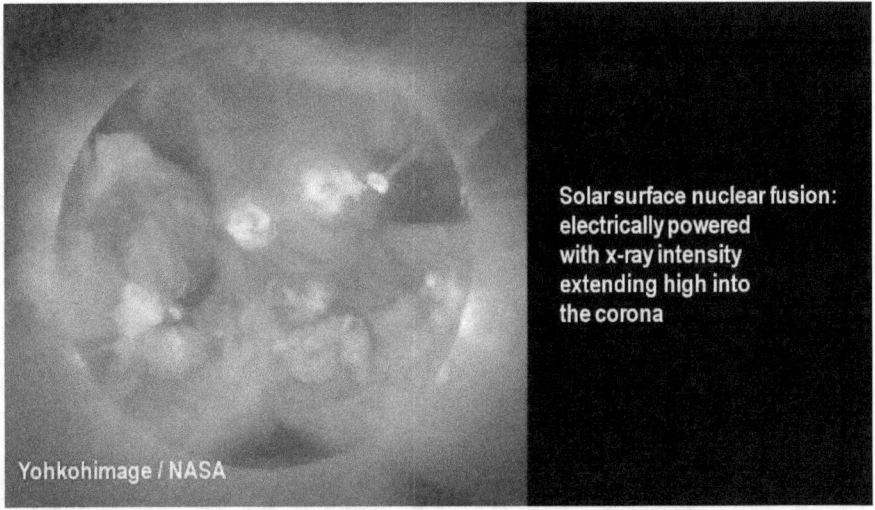

Solar surface nuclear fusion: electrically powered with x-ray intensity extending high into the corona

Yohkohimage / NASA

So, let's explore some of the details of the electric evidence that we see everywhere in the Universe, which our future depends on, including our energy future as we avail ourselves of the nuclear-fusion process of the universe that synthesizes atomic elements with electric plasma fusion at the surface of every Sun.

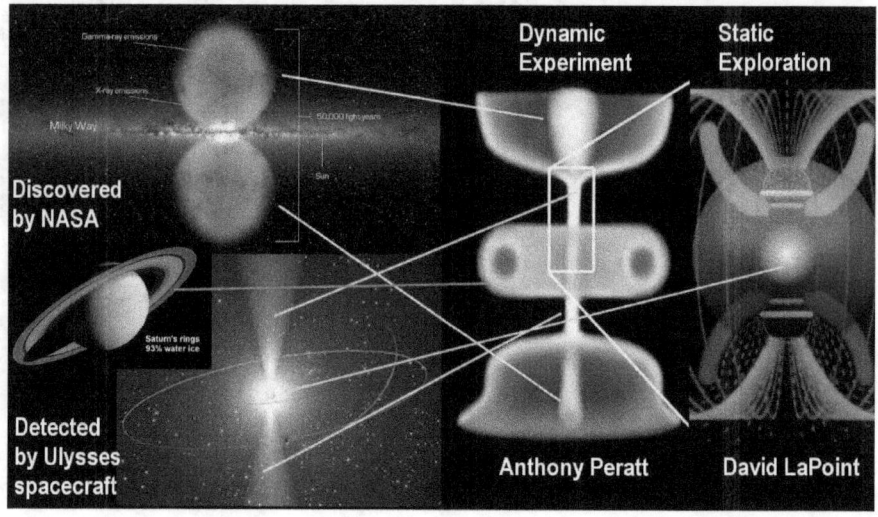

** The Common Electric Foundation, that is universally evident

As I already said, the common platform for the Galactic System, the
Solar System, and the Planetary System, is electric in nature. The
supporting evidence is found in the fact that major plasma-electric
features are expressed on the galactic level as extensively as it is on
the level of the solar system, which in addition can be replicated in
the very small in the laboratory in the form of electric physical
experiments.

As I have also already indicated, in laboratory experiments the
expression of electric principles have resulted in the discovery of
geometric phenomena in plasma-flow dynamics that are unique to
electromagnetic phenomena. The presence of aspects of the unique
geometry in the galactic and solar systems indicates what type of
cosmos we live in. As I already indicated, too, we see the unique
electric geometries surprisingly widely reflected in both the large
structures of galaxies and in the smaller structures of solar systems,
and in the still smaller planetary systems, such as Saturn's systems
of rings and moons. While only specific elements of the complete

experimentally derived geometry are detected in these examples, the corresponding nature of the evidence indicate that both the galactic system, and the solar system, and even the planetary systems within them, all operate on the same basic electromagnetic platform. It would be surprising if this wasn't the case.

Tell-tale features of plasma-electric dynamics

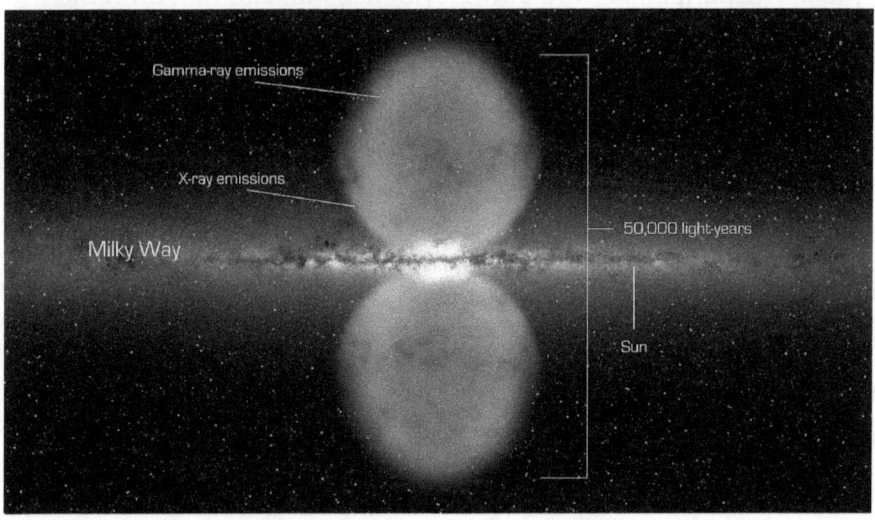

As I said before, one of the tell-tale features of plasma-electric dynamics of the galactic system has recently been discovered by NASA in x-ray and gamma-ray light, in the form of two gigantic plasma confinement structures that extend above and below the galaxy perpendicularly to the galactic disc, extending outwards from it across 25,000 light years in either direction. But where does the plasma come from that is magnetically concentrated under these two large confinement domes?

Electricity in space

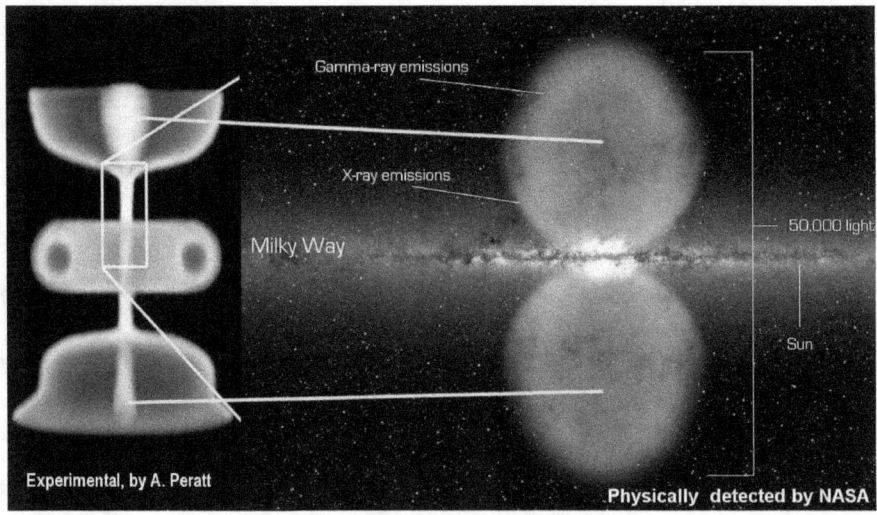

The question for the plasma origin can only be answered by correlating what we see with the known geometry that has been discovered in high-power laboratory plasma experiments where the dynamics has been made visible, artificially.

Electricity in space is known to be carried in the form of plasma particles that are invisible because of their small size. The larger of the plasma particles are 100,000 times smaller than the smallest atom. However, the effect of plasma, which is nearly always in motion, becomes visible by its interaction with atomic structures that alone are able to emit light, which happens in space when their agitation by flowing plasma is strong enough.

In the galactic system, only the highly concentrated plasma under the confinement domes has become visible. This does not mean that the remaining parts of the Primer Fields of the galactic system do not exist. It only means that the task falls on us as human beings to complete in the mind the galactic system by including the unseen on the basis of scientifically discovered principles.

The magnetic pinch effect that shapes intergalactic plasma streams

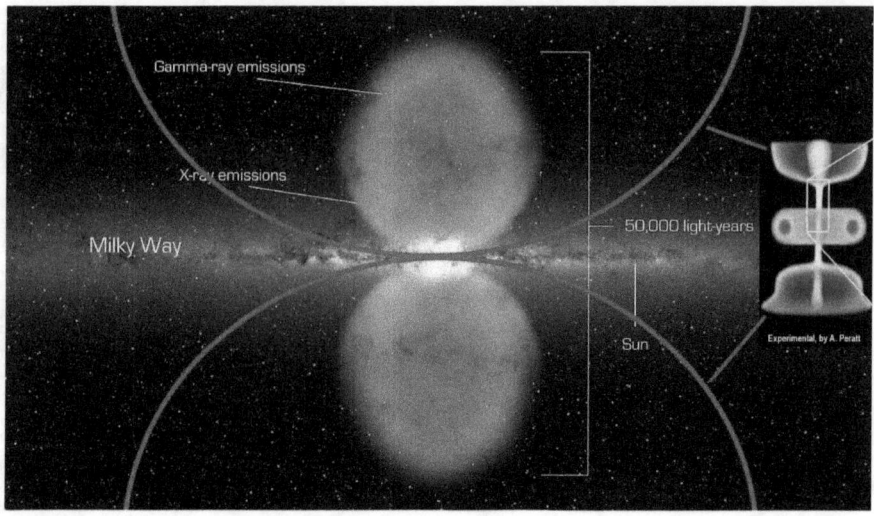

This means that the magnetic pinch effect that shapes intergalactic plasma streams into ever-tighter confinement remains nevertheless too weak to be visible. Only when the plasma confinement becomes immensely dense, perhaps a thousand-fold, under the confinement dome, do traces of the plasma system become visible, which thereby stand as proof for the whole.

By including in the mind the inherently invisible

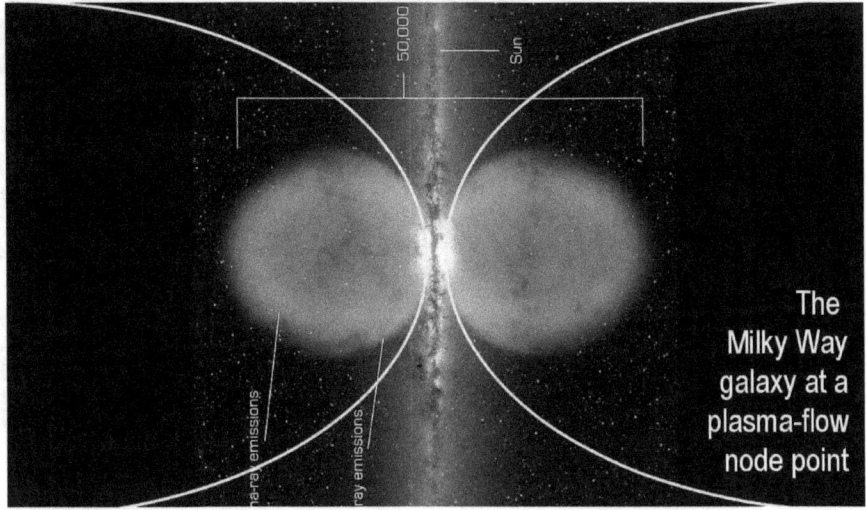

The Milky Way galaxy at a plasma-flow node point

By including in the mind the parts of the galactic system that are inherently invisible, the galactic disk comes to light as being located at the node point of intergalactic plasma streams. Since plasma-flow dynamics has been extensively explored in laboratory experiments, in recent times, it becomes but a small step in the mind to complete the geometry that must exist for the visible part of the plasma system under the confinement dome to exist, which is visible in x-ray and gamma-ray light.

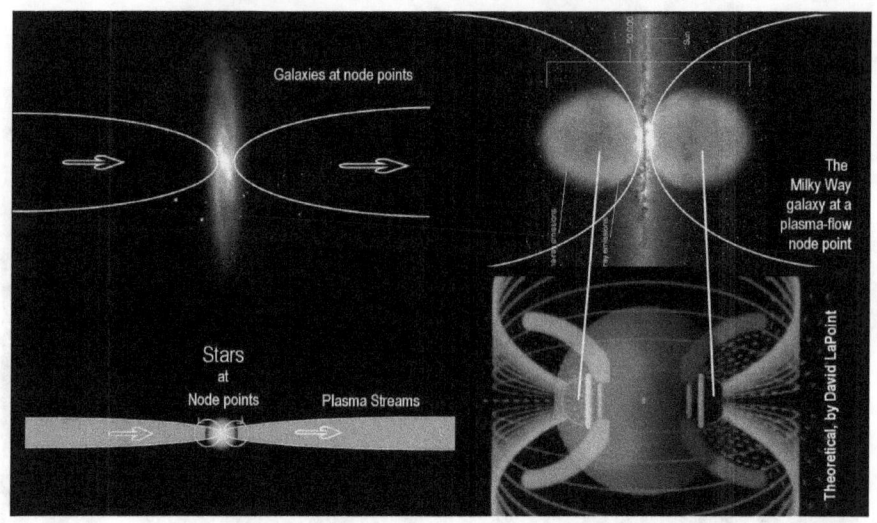

Galaxies at node points

The Milky Way galaxy at a plasma-flow node point

Stars at Node points

Plasma Streams

Theoretical, by David LaPoint

This means that in real terms strong evidence exists that all galaxies are located at node-points of intergalactic plasma streams. We see this evidence rather strikingly when we extend our vision wide enough to encompass many galaxies at once.

The Capodemonte deep field of 35,000 galaxies

In this partial view of the Capodemonte deep field of 35,000 galaxies, we encounter many galaxies lined up in short and long strings in this flattened image of space. By seeing their string-like arrangements, we see in the mind the existence of intergalactic plasma streams. Thus we behold in the mind, once again, what is physically invisible, but which we know must exist for the galactic system to function in a manner that incorporates the visible parts of the evidence.

Thus, here too, the human mind that is capable to see scientifically the physically invisible, fills in the blanks and completes the structure according to scientifically known dynamics.

By the Nobel Price Laureate Hannes Alfven

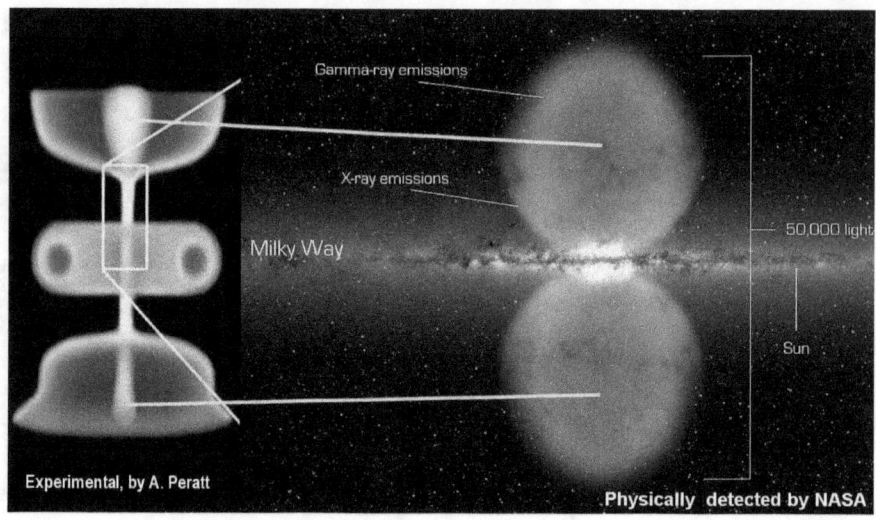

At this stage a new question comes to the fore-front. What happens to the concentrated plasma between the confinement domes, in the galactic system? The experimental evidence indicates that a portion of the energy flowing through the system dissipates laterally, that is perpendicularly to the main flow. In the galactic system, the lateral plasma flow, necessarily extends across the entire galactic disk.

The distribution of the plasma within the galaxy occurs in the form of magnetically self-confined Birkeland currents that are magnetically twisted and rotating, as shown in the model for the galactic system presented by the Nobel Price Laureate Hannes Alfven

Only a small portion of the plasma that flows in the interstellar and intergalactic streams is consumed at the node point where the galaxy is located. Most of the plasma flows on towards to the next node point. Being weaker now, the out-flowing current expands as it flows away with the electromagnetic pinch effect being reversed. The entire galactic node point becomes thereby a complex structure of interacting electromagnetic fields, without which the high-density plasma compression cannot be achieved that is required to power a sun within a solar system, or on the larger scale is required to power a galaxy of suns, as each star is a solar system.

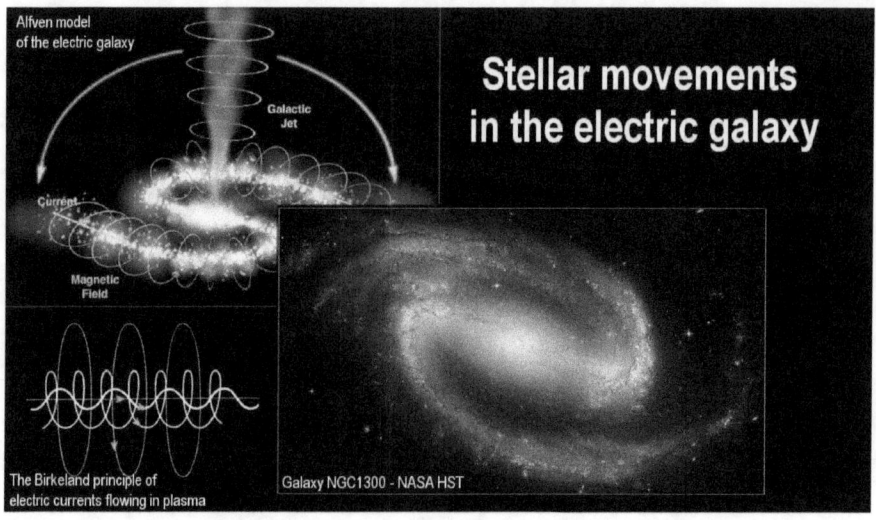

One of the very early models of the electric galactic system, is the Alfven Model, named after the 1970 Nobel Price recipient Hannes Alfven. He saw the galactic spiral arms as plasma conduits that are magnetically self-aligned, and rotating axially within them according to the principles of the Birkeland currents. While the Alfven Model fails to accommodate some of the modern evidence that was not known at the time, Alfven's basic platform is nevertheless sound, which renders the spiral arms as flowing plasma currents that are drawn into axial rotations within them, from which the stars derive their motion that are falsely recognized as orbital motions. Gravitationally bound orbits of stars is totally impossible on the galactic scale, according to the known laws of orbital dynamics. Only electromagnetically caused motions are physically possible in a galactic system. These includes both the axially rotating motions of stars in the Birkeland currents of the spiral arms, and the rotating motion of the entire interlocked galactic disk spinning by the electromagnetic effects of the galactic primer fields.

Black Holes
Under
the Stars

** Dark Age of the Black Hole

The theory of the Black Hole in space has become a prison for astronomy in the form of a doctrine that shuts out 99.999% of the Universe that the doctrine doesn't allow to be recognized. Under the doctrine, plasma in space, which has been recognized in plasma physics to constitute 99.999% of the mass of the Universe, is denied in astronomy, to even exist.

Of course, with so much of reality denied, astronomy finds itself seriously challenged when it comes to understanding phenomena of plasma physics that are deemed not to exist. To solve the impasse, 'modern' astronomy has invented a method of self-deception in the form of theorized black holes in space that are deemed to have such an immensely compacted gravitational force that even light cannot escape from it, hence the term, Black Hole. It is further theorized that whatever light exists in surrounding space is distorted by the hole's immense gravity and is channelled around it by the theorized effect of gravitational lensing.

In real terms the Black Hole doctrines act as a trap

In real terms the Black Hole doctrines act as a trap that enables astronomers to dream about the Universe in simplistic terms without having to bother facing reality. It doesn't seem to matter in the thereby created constricted box of Old World astronomy that the phenomenon of hyper-dense black holes is physically impossible. Black holes are deemed to be the remains of expired stars that had their atoms ripped apart by the force of gravity in an explosion that left nothing but neutrons remaining, which, lacking the repulsive electric force of the protons, can form a lump in space with ten times the mass of our Sun, compressed into a sphere of just 30 kilometers.

The problem that makes such an exotic theory impossible, is the well known fact in atomic physics, that neutrons are only able to exist within the dynamic sphere of atomic structures. When neutrons are split off from atomic structures, such as in nuclear reactors, neutrons quickly decay back into protons and regain thereby their original positive electric potential. The fact that the theorized phenomena are not possible, points the finger back at the plasma universe where extremely high density concentrations in plasma are possible

The same can be said about the super massive black holes that are theorized to exist at the center of galaxies with a mass 100 million times that of our sun, compressed into a sphere up to 60 billion kilometers wide.

Believed to be a super massive black hole

An unusually bright X-Ray flare from Sagittarius A, deemed to originate from a super-massive black hole, 4.3 million times as massive as the Sun, located near the center of our galaxy, seen on January 5, 2015.

detected by NASA's Chandra X-ray Observatory

Artist conception of a black hole

What is believed to be a super massive black hole has recently been discovered near the center of our galaxy by a strong x-ray emission. The black hole is believed to contain a mass 4.3 million times that of our Sun, contained within a sphere 18 billion kilometers wide.

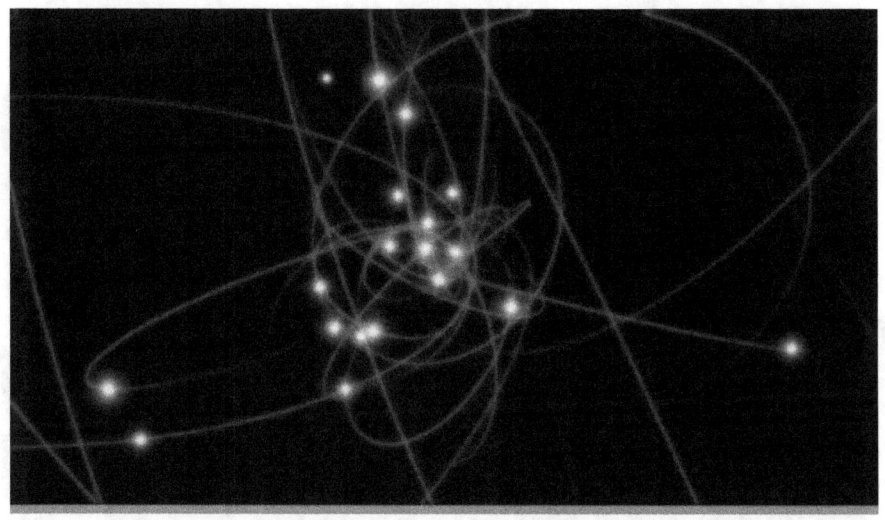

In visible light, the black hole cannot be seen directly>, but can be detected by its strong apparent gravitational effect on a group of stars that surround it. `The group of stars shown here has been observed for more than a decade with the European Southern Observatory's Very Large Telescope.>

From the changing positions of the stars

A super-massive
plasma shpere
4 million times the mass of the Sun
100 times larger than our Sun

2011

From the changing positions of the stars, their potential orbit has been calculated according to the laws of gravity bound orbital motions discovered by the German astronomer Johannes Kepler in the early 1600s. The position of a large concentration of mass is indicated by the potential orbits of the stars around the theorized central mass. The calculated orbits all match this pattern. But what specifically is this large central mass that all observed stars orbit around? What kind of plasma phenomenon can have the observed effect? A plasma stream 100 times as wide as the Sun, could theoretically contain this volume of mass in the form of plasma.

The observed massive object may be 18 billion kilometres wide

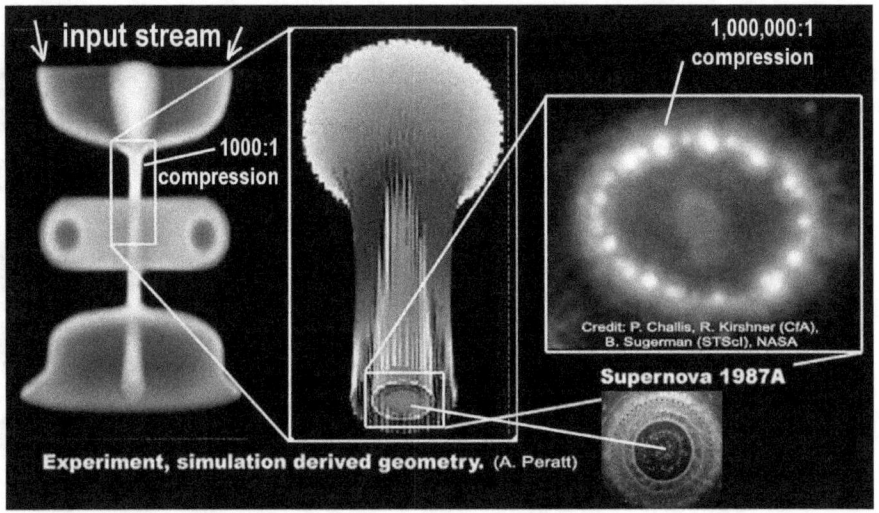

Experiment, simulation derived geometry. (A. Peratt)

However, laboratory experiments have shown that high-energy plasma streams tend to organize themselves into a circular ring of filaments. For this reason, a much larger plasma stream is required to weigh in with the effective gravitational effect of 4.3 million suns. Astronomers have calculated that the observed massive object may be 18 billion kilometres wide, or roughly 13,000 times as wide as our Sun. If so, the result would match in principle what one would expect to see for a large plasma current in a galactic system between primer fields.

If one applies the laboratory experiment as a model to the galactic system, the orbiting stars and the theorized central mass, perfectly match the experimental model.

Stars are spheres of plasma

A deemed star-forming region in the Large Magellanic Cloud. NASA/ESA image

The observed orbiting stars would be located in this plasma-system's star-forming region. Stars are spheres of plasma. They begin their existence as fields of dense plasma concentration within the electromagnetic system of the primer fields that interface the galactic bulge with the spiral arms. The plasma fields grow in this environment and gradually condense by the force of their internal gravity.

As the gravitational pressure builds up

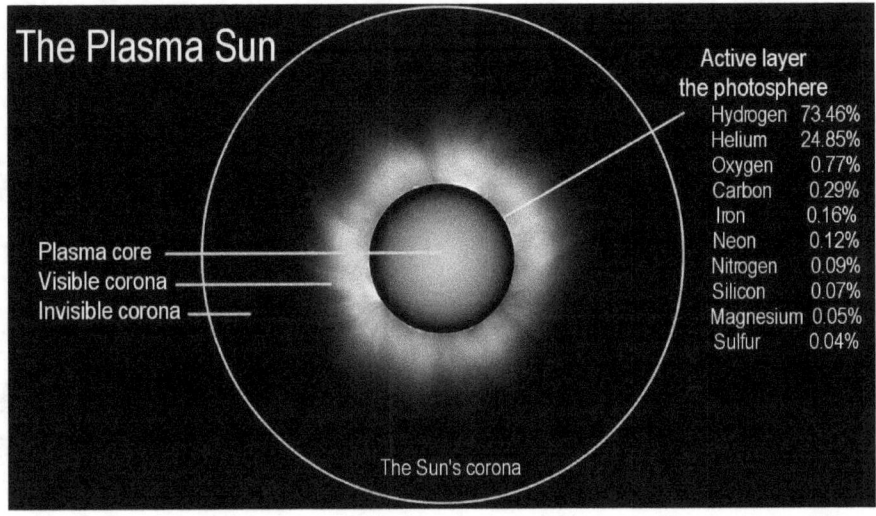

As the gravitational pressure builds up in the condensing plasma fields, the electrons in the plasma mix tend to migrate to the outer layers of less gravitational plasma pressure. Eventually a stage develops where electric interaction occurs between the resulting electron sheet and the overlaying plasma. At this point plasma fusion reactions begin to happen, by which in effect, a sun is born. The fusion reactions fuse plasma into electrically neutral atomic elements, which tend to flow away with the solar wind. By this process plasma is consumed and atomic elements for the planets are created. The entire process is sustained by the fusion reactions consuming plasma, which motivates a large stream of plasma flowing into the now active sun, or star, as a distant sun is called. In a galactic system that powers 200 to 400 billion stars, substantial plasma streams are drawn from intergalactic space to power the process. Likewise, substantial volumes of plasma flow into the galactic spiral arms.

Extremely large plasma streams are formed

A deemed star-forming region in the Large Magellanic Cloud, NASA/ESA image

This means that at the interface between the galactic bulge, and the spiral arms, extremely large plasma streams are formed by the physics of their node points. The plasma concentration of the node points at this interface is so extremely large that its gravitational effect on the nearby stars can be recognized. The phenomenon is erroneously believed to be caused by a super massive black hole, which in reality is merely a node-point plasma stream. According to the calculated values, the detected plasma stream in this region, termed a black hole, is a ring of plasma 18 billion kilometers wide. A super massive "black hole" is not invisible in space for the reason that its gravity is so great that no light can escape from it. Instead it is invisible, simply because pure plasma in space, no matter how massive it may be, emits no light. Light is emitted only when a plasma stream is polluted with atomic elements in its path.

A moving gas cloud has been discovered

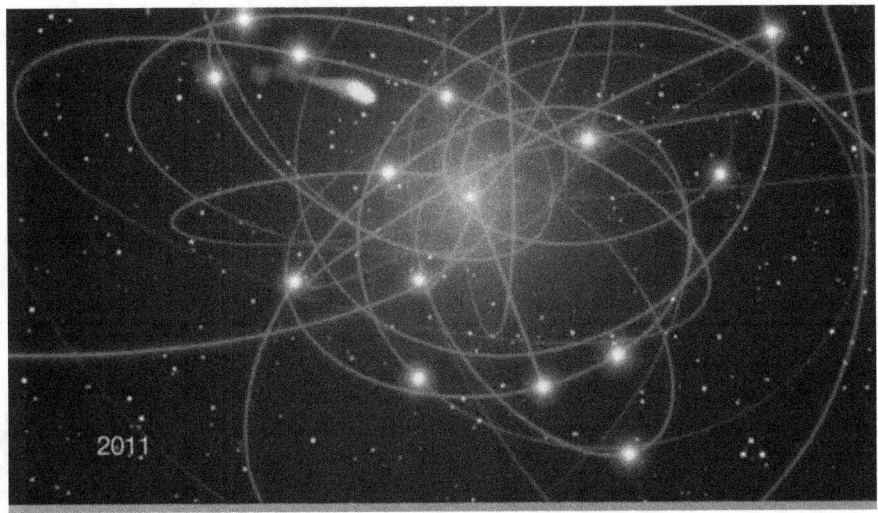

In the field of stars that is observed by the European Southern Observatory with its Very Large Telescope, a moving gas cloud has been discovered. The cloud may have originated as a gas planet that became lost from the emerging solar systems. It glows brightly in the observed plasma in the region. This may also be the object that the Chandra Observatory saw as an x-ray flair. A plasma stream emits no light, x-rays, or gamma-rays. Only the interaction of plasma with the ordered structures of atomic elements has this capability.

When 99.999% of the Universe is omitted

A simulation of the future of a gas cloud that has been observed approaching the supermassive black hole at the centre of the galaxy, with the expected positions of the stars and gas cloud projected for the year 2021" by ESO/MPE/Marc Schartmann - www.eso.org/public/images/eso1151a/. Licensed under CC BY 4.0 via Wikimedia Commons

It is assumed that the observed gas cloud will eventually be torn to shreds by the gravitational pull of the so-called "black hole." It is more likely that the gas-cloud will simply evaporate and flow away with the "wind" of the plasma stream, just as the new-born planets themselves will eventually be drawn into the spiral arms. The paths of the orbits that have been added to the image have not actually been observed, but have been mathematically projected, based on Kepler's law of gravitational orbital motions. The real motions will likely turn out to be totally different as the planets move in a whirlpool of plasma that comes with its own characteristics. When 99.999% of the Universe is omitted, the perception of it becomes flawed. That the whirlpool of plasma in this region does exist, and is extremely dense, is evident by the intense agitation the gas cloud experiences on its path, so that it even glows in x-ray light.
So, what do we see here, in the field observed by the ESO team?
We see a tiny region of potentially a great number of such places, where stars are being born. The observed region is roughly 50 light days wide. This makes it 30 times larger than the width of our solar

system, a small speck indeed. The detected "black hole" within, is deemed to be roughly 7/10th of a light day wide, which it may and may not be, as a plasma node. This means that what is shown here, in the image produced by ESO, as large as it may be, captures just a tiny region within the large galactic disk system that is recognized to be 2000 light years thick and may extend for 180,000 light years laterally.

Processes going on in our galactic system

This means that what we see here is merely an example of countless such processes going on in our galactic system.

Thousands of super massive node points

galaxy Messier 83

NASA, ESA, and the Hubble Heritage Team

There may be thousands of similar super massive node points in thousands of plasma streams operating all-the-way around the central bulge of our galaxy, and many times more so in the larger galaxies, all interfacing the galactic spiral arms with the main intergalactic plasma streams. The so-called "dust lanes" are congregations of atomic elements that escaped from the operating billions of solar systems and flow along with the plasma currents in the spiral arms.

The combination of experiments

The combination of experiments, observations, and scientific honesty takes the mystery out of the Black Holes philosophy and clears the field of exotic dreamscapes.

Spinning fantastic yarn around black holes

As sciences advances, it becomes less frequent that respected names in astronomy stand before the world still spinning fantastic yarn around black holes that either do not actually exist by being physically impossible as the tales are told, or exist in the form of plasma structures that can be replicated in laboratories, but which the doctrines insist are not possible as plasma streams are not recognized to even exist.

Hans Christian Anderson tale of the Emperor's New Clothes

The tale of the
Emperor's New Clothes

by Hans Christian Andersen in 1837
ilustrated by Vilhelm Pedersen (1820 - 1859)

The resulting sad situation reflects what Hans Christian Anderson had illustrated long ago with his tale of the Emperor's New Clothes, where the Emperor parades himself through the streets with clothes produced from fabric woven on empty looms, which all proclaim to be beautiful. The collective deception that society had bowed to, had worked amazingly well in the tale, until a child proclaimed the obvious truth, namely, that the parading Emperor is naked.

the Flat Earth doctrine fell by the wayside

The Flat-Earth Doctrine

Engraving by Flammarion (1888)
of a traveler at the edge of a flat Earth,
who sticks his head through the firmament.

It appears that the Flat Earth doctrine fell by the wayside on this type of path of scientific honesty some time ago. The current Dark Age of the Black Holes doctrines of all sorts, which also includes the doctrine of the internally heated sun, may end some day in a similar manner.

Society is still trapped into the Anderson tale

The tale of the
Emperor's New Clothes

by Hans Christian Andersen in 1837
ilustrated by Vilhelm Pedersen (1820 - 1859)

The obvious fact that society is still trapped into the Anderson tale, and similar tales, and not only in astronomy, but also in nearly all areas of civilization, should inspire an exploration on the wider front, to separate what is deemed practical, from what is demonstrably efficient. The Universe, certainly is extremely efficient in all its aspects, otherwise it would not exist. Maybe we should take the Universe as an example and make an effort in ourselves to step away from the small-minded practicality that society has become accustomed to, for far too long.

The Efficient versus the Practical the Old-World concept of the Sun

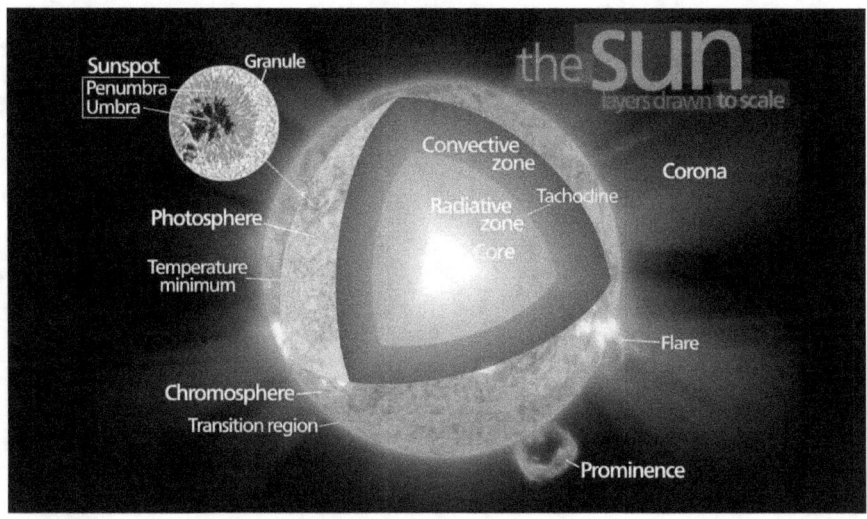

** The Efficient versus the Practical

In common perception the Old-World concept of the isolated, internally heated Sun, is deemed highly practical, isn't it? Here the Sun is a sphere of hydrogen gas that is compressed by gravity and heated to millions of degrees, which causes hydrogen to fuse into helium. The concept is so simple and practical that one can trust it to be true, without one having to explore exotic universal principles, and effects that no one can actually see. The internal heated sun theory is so simple that it is hailed by the simple minded. It is so simple that one can describe the universe with it in 3 easy sentences, and move on from there to better things. That's being practical, isn't it? In comparison, the plasma Universe, with its electrically powered plasma Sun, is deemed totally impractical. It is deemed so impractical that no one even tries to understand it.

The plasma Universe and the plasma Sun are impractical

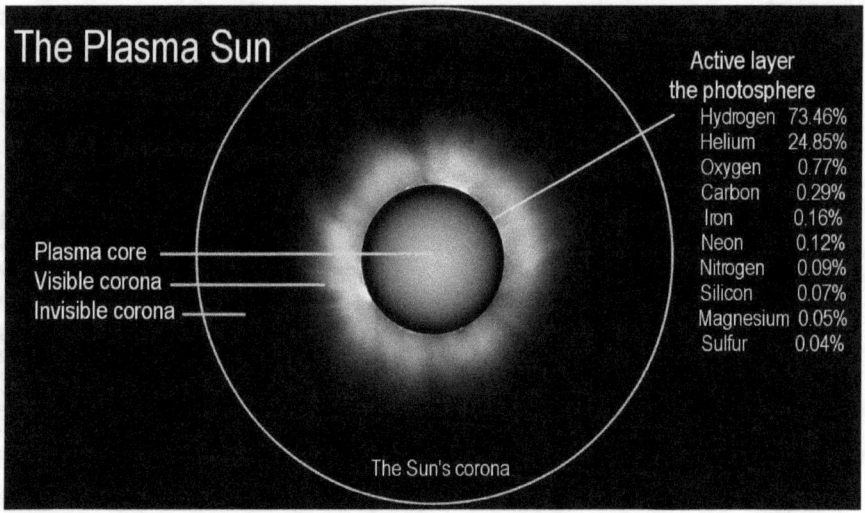

Indeed, the plasma Universe and the plasma Sun are impractical concepts that require one to get out of the easy chair and to focus on what is efficiently real. The Universe is not simple in this respect, but it is efficient. It is so efficient that it actually functions, while the Old Sun concept is a nice tale that's physically impossible.

*The old tale is so impossible

Joint European Torus

wikipedia

The old tale is so impossible that the world's science community has not been able to replicate the theorized high temperature internal fusion-power process that is deemed to power the Sun. After several decades of laboratory experiments in many countries in the world, none have succeeded to replicate the internal heating of the Sun, for the obvious reason that such a process doesn't happen in the Sun. In every experiment the fusion product diluted the fusion fuel and blew the process out. The world record established by the Joint European Tokomak experiment, is a fusion burn of slightly less than a second, or slightly more than a second at half power. The failure illustrates that the world aims to replicate a solar process that actually isn't happening in a sun. That's what is called the leading edge of practicality.

A sun could never operate on this self-defeating basis

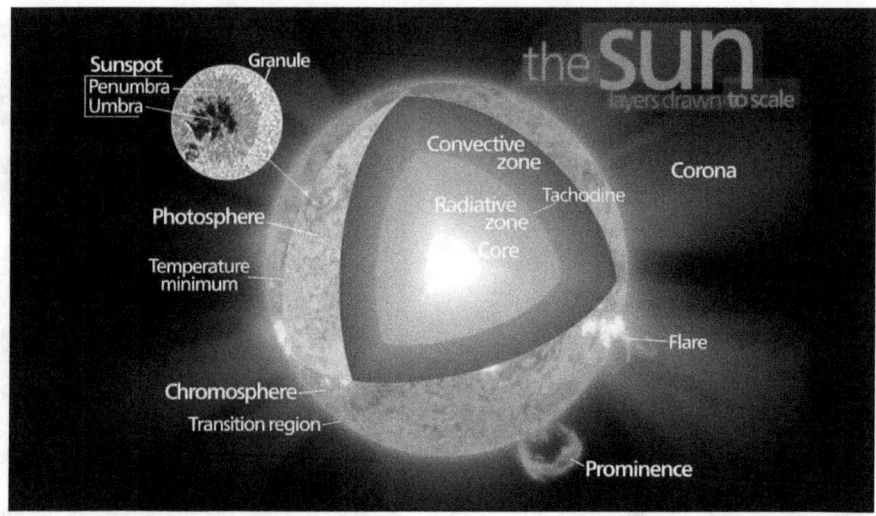

A sun could never operate on this self-defeating basis. Its fusion core would have been clogged up with its fusion product, eons ago. The heavy atom of helium would have congregated at the core and blown the Sun out forever. Obviously this hasn't happened.

The Sun operates on a platform that actually functions

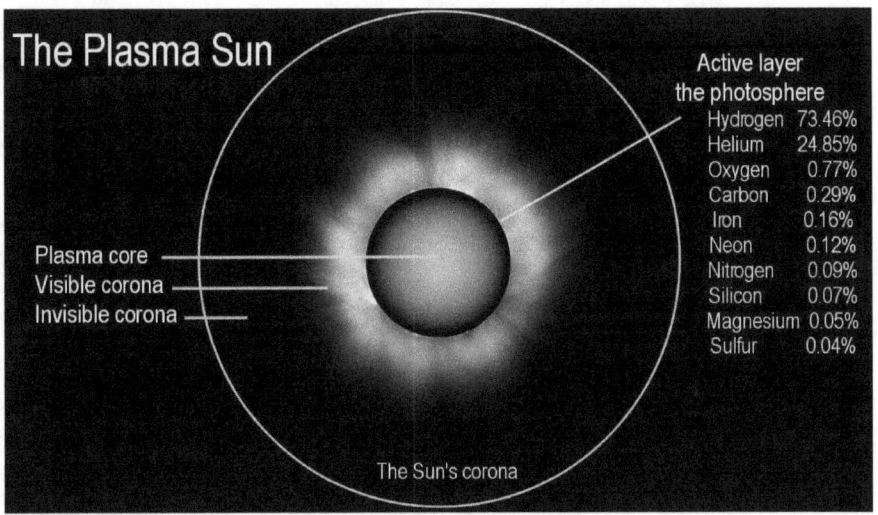

The Plasma Sun

Plasma core
Visible corona
Invisible corona

The Sun's corona

Active layer
the photosphere

Hydrogen	73.46%
Helium	24.85%
Oxygen	0.77%
Carbon	0.29%
Iron	0.16%
Neon	0.12%
Nitrogen	0.09%
Silicon	0.07%
Magnesium	0.05%
Sulfur	0.04%

It is plain to see that the Sun operates on a platform that actually functions, that is so efficient in its design that the atomic elements that are produced in the fusion process on the surface of the Sun, are dynamically vented into space in the flow of the solar wind and accumulate on the planets. That's efficient. The entire solar system is created on this basis in a process that still continues. That's creative efficiency.

Link between the Eurasian Continent and the American Continent

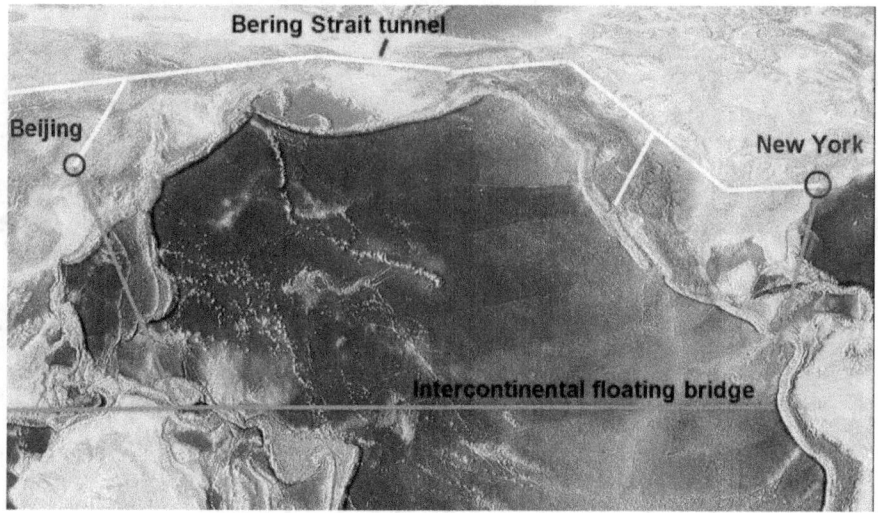

While the solar comparison is extreme, let me illustrate the critical difference between the efficient and the practical, with a less extreme example.

Consider the case of implementing the long-desired railway transportation link between the Eurasian Continent and the American Continent. How would one create such a link that would bridge the isolation of the two continents and their people?

Construct a 100-kilometre-long tunnel under the Bering Strait

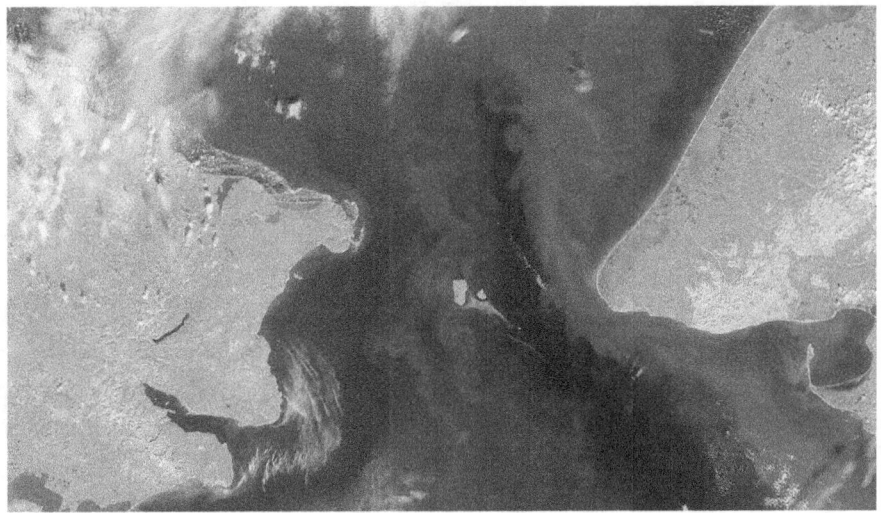

One way to create such a link, would be to construct a 100-kilometre-long tunnel under the Bering Strait between Alaska and Siberia, and to lay down railway lines to the far North on both continents, to link up with the tunnel. The building of the project would be a wonderfully practical option. Wouldn't it be? Putting the spade into the ground, so to speak, for such a project is totally practical, isn't it?
As a national effort on both sides, the construction could be completed in a dozen years. Hundreds of trains could then make the journey each day. The link could serve to economically develop the regions in the far north near the big trunk line. All of this can be done.

All of it is totally practical

DEW line station at Point Lay, Alaska.

All of it is totally practical, even though the resulting long and slow howl over the top of the world in the barren landscape that only the military finds useful, often at sub-zero temperatures, especially in the winter, and much of it crossing difficult terrain, might add up to a rather long journey of many days, if not weeks. But it is all practical to do.

A 10,000 kilometre floating bridge

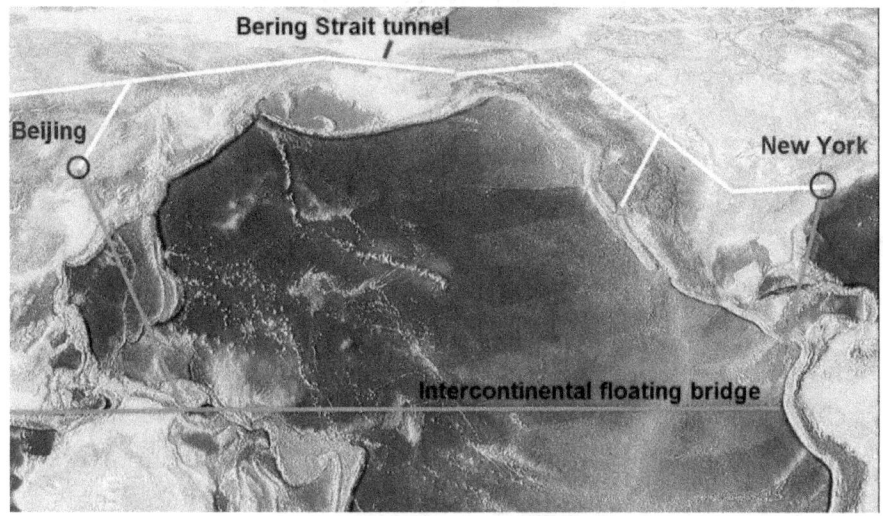

In contrast, another option for meeting the stated objective to connect the two continents with a fixed link, would be to lay down a 10,000-kilometre-long floating bridge across the Pacific, such as from Shanghai to Los Angeles in California. Such an immense project would be considered highly impractical, right? Indeed, it would be that, by the definition of what is practical. But it would be efficient.

Consider that the entire sea-bridge could be produced in large scale automated industrial processes, utilizing nuclear power and heat-shaped basalt to create the bridge modules. On this automated basis the bridge could be created with very little laborious effort, and in a relatively short time. Of course, one would have to create the automated industrial infrastructure for it, in order to facilitate the construction process. But once the infrastructure is up and running, it would require little human labour to produce the floating bridge. The perfectly level sea bridge would also enable high-efficiency mass transport, and very-high-speed passenger service. It would enable trains to be placed into vacuum channels for speeds exceeding 10,000 miles an hour. The transit time across the Pacific would thereby be reduced to less than an hour. In addition, the sea bridge would serve as a trunk line for a completely new type of agriculture, laid afloat onto the sea. Networks of floating agriculture would be extending outwards from the sea bridge for as far as the need for more food would require it. It would be constructed together with thousands of floating cities along the way to service the agriculture. The vast benefit in every direction would make the project efficient.

Since the automated bridge production process for this type of infrastructure is inherently immensely efficient, the revolutionary freedom in physical production that such a process enables for mankind, would subsequently be applied to all kinds of industrial processes that are not even considered yet, as for example the mass production of complete housing units that cam be manufactured with such a great efficiency that society can provide high-quality housing to each other for free as an investment into itself.

Quality housing is a critical component

Quality housing is after all a critical component for raising the creative and productive potential of society on the recognition that the human being is the most precious asset that a society has, being the measure for everything that has true value. The striving at the leading edge in civilization to advance human development to the utmost, is the most efficient path in civilization to increase the quality and the power of society's living. Practical considerations would close the door to all that. We need to step past the sad limits of perceived practicality.

The sea bridge is efficient

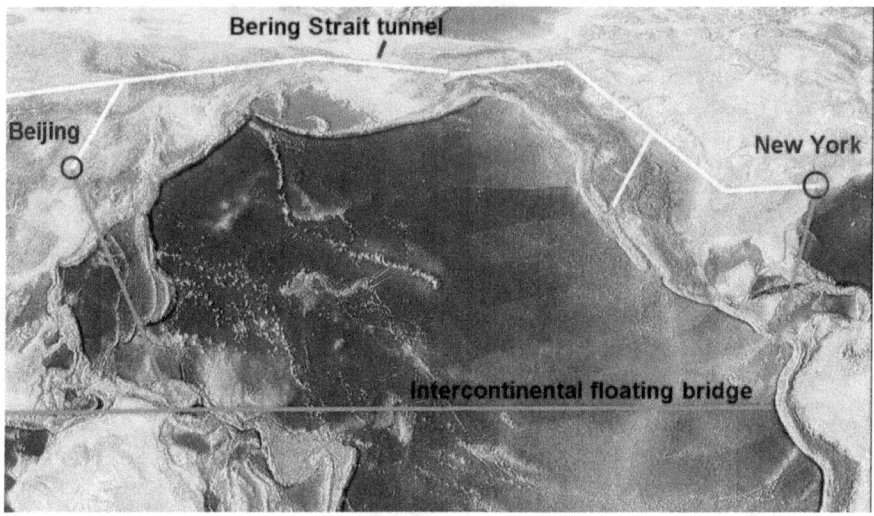

Sure, the very concept of an intercontinental sea bridge is considered impractical, even impossible. Indeed, who would pay for such a massive effort? In the current world of small-minded thinking, nothing is deemed practical unless one has the money to fund the effort, in addition to funding universal swindlery, profiteering, looting, and monetarist gambling orgies that have imprisoned entire nations into debt-bound concentration camps with strangled economies. In this landscape of what is deemed practical, nothing can be built for the well-being of society. This consideration renders the intercontinental sea bridge, totally efficient. The sea bridge is efficient, because, contrary to all practical considerations, it would raise humanity's power to live, to new levels of freedom and self-development.

The practical world has become so small

Annihilation is assured

500,000 times
Hiroshima
in one hour

Castle Bravo - the first U.S. test of a dry fuel thermonuclear hydrogen bomb - March 1, 1954 at Bikini Atoll, Marshall Islands

In comparison, what is limited to being patently practical is inherently destructive, as it denies the power and the nature of the human being. The small-minded practicality is literally destroying humanity. It has destroyed society so deeply that nuclear war is accepted as a reality that has no solution while the clock is ticking that has brought the world closer to nuclear war than at any time in the past.

The practical world has become so small that the masters of empire now dare to demand that 6 billion people, out of the 7 billion world population today, be killed by famine, diseases, war, and other means of genocide, while society supports the insanity. That's the face of what is deemed practical.

The equatorial sea bridge

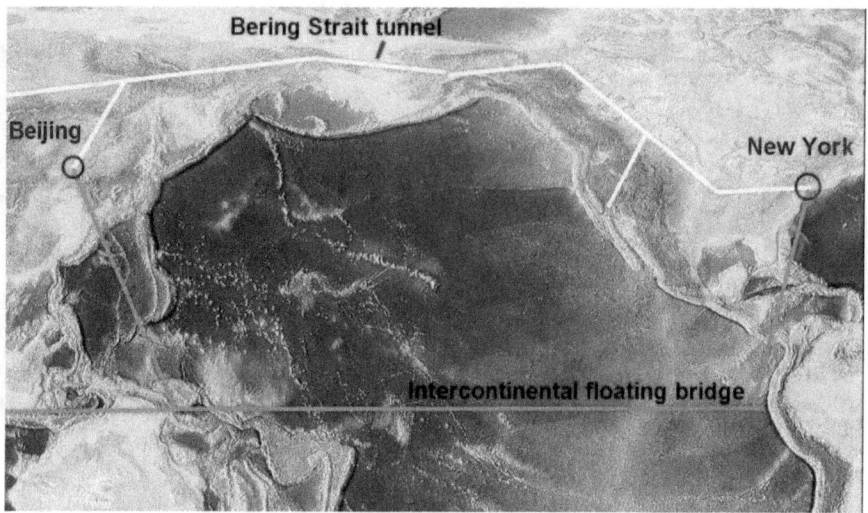

Bering Strait tunnel

Beijing

New York

Intercontinental floating bridge

Now extend the comparison still further. Let's choose an even more-efficient route for the sea bridge. Let's take it all the way south to the equator where we need to have much of the world's agriculture located in 30 years when the next Ice Age begins with a 70% weaker Sun and disables agriculture in all regions outside the tropics. The equatorial sea bridge would be significantly longer, of course, but it would be immensely efficient in that it would enable humanity to continue to live under the dimmer Sun in a colder world. That's the measure of peak efficiency.

The Glass Steagall system of legislation

1933 - the Glass-Steagall law to protect that value of the U.S. currency - now repealed

Another example of an efficient system, this time in politics, is a project that takes us to a still-higher level far above the level of the small-minded practical concerns. This project is the restoration of the Glass Steagall system of legislation that is designed to purge the fabric of civilization of the infestation brought upon it by the forces of empire that are well known by their ugly hallmarks.

A debt-bound concentration-camp system

"The City" in London, one of the largest financial centres in the world

The hallmarks of the dying world are swindlery, profit looting, and monetarist gambling orgies, and so on, which have imprisoned entire nations into a debt-bound concentration-camp system that strangles the dying economies where nothing works anymore that supports civilization, so that people die in large numbers from the consequences as they once had in the Nazi concentration camps that some may still remember.

The criminality in the private camp system

Capitol Hill
Washington DC - USA
wikipedia

The criminality in the private camp system is so extensive and so deep-reaching that the highest ranking justice official in the USA has publically announced to the world that the extreme criminality cannot be prosecuted, as any attempt to do so would immediately collapse the entire western financial and economic system from the Euro to the Pound to the Dollar. The speech implies that the criminal system is rotten to its very core with no hope remaining for its redemption, so that the national justice system has simply capitulated before it.

Practicality of the evermore deadly economic criminality

Thus, the hailed practicality of the evermore deadly economic criminality that is increasingly supported with floods of public bail-out money that the public takes on as debt, has its deep-core criminality officially legalized from the highest ranks of government. And this is deemed practical. It is as practical as a march of the blind-folded over the proverbial cliff, which it actually is.

Greece a nation of 11 million people

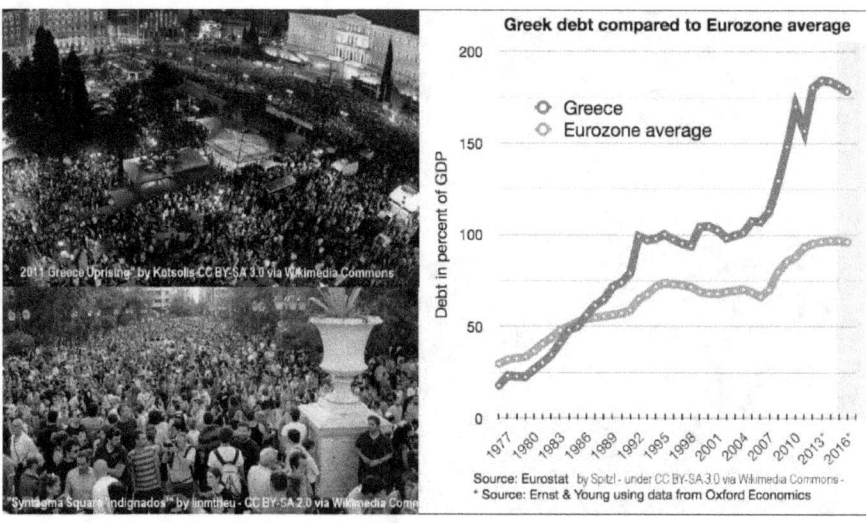

Greek debt compared to Eurozone average

○ Greece
◇ Eurozone average

Source: Eurostat by Spitzl - under CC BY-SA 3.0 via Wikimedia Commons -
* Source: Ernst & Young using data from Oxford Economics

Greece is an example of a nation trapped into the choking practicality, the practicality that has turned the Euro Zone effectively into a concentration camp where society is imprisoned without hope in sight. Greece is a nation of 11 million people. Almost 300 billion Euros in debt has been forced onto the nation to bailout the banks of the camps system. In the bail-out process the swindlers claim the gold, and the people pay for it with having their life decimated and destroyed. Each person now owes a debt of 32,000 Euros for loans that went directly to bailout the banks that now choke the nation to death with interest payment demand and fascist conditionality.

Glass Steagall protected America

1933 - the Glass-Steagall law to protect that value of the U.S. currency - now repealed

The Glass Steagall system of legislation that had in its time protected America and the western world, but which was repealed, remains still standing in principle behind the horrid scene to be re-applied at the present time to serve as an efficient system of regulatory means for the urgent task of closing the private imperial camp system down.

Sure, Glass Steagall isn't practical for the thieves as it doesn't allow any attempts to reform the highly deadly camp system into a less deadly camp system. Glass Steagall stands as an efficient means for protecting society and civilization against all forms of financial criminality, new or old, and to restore thereby the needed freedom of humanity to pursue its self-development and its happiness. Glass Steagall effectively rips the practicality out of the hands of the masters of war and depopulation - the two deadliest of the hailed flagships of the camp system of empire.

There is nothing small about the Universe

strings of galaxies and stars

ESO/VIMOS galaxy cluster ACO 3341

ESA/Hubble & NASA
Acknowledgement: Claude Cornen

In the small-minded sense, where practicality has nothing to do with reality, the Universe stands as totally impractical. It operates on a higher level than practicality. There is nothing small about the Universe, weak, isolated, disconnected, or running down. The Universe is extremely efficient in all its aspects. The concept of practicality is far out of sight. Instead, the Universe is light, which means, it functions immensely.

Life itself, is not practical either, in the modern sense. Nothing about life is small. Life is efficient, big, profound, dynamic, progressive, expansive, forever new and evermore grand.

Life is an 'image' of the nature of the Universe

Life is an 'image' of the nature of the Universe. It defines the New World that beckons us to leave the old world behind. Life is God, so to speak. God is not practical. God is big, real, amazing. There is nothing small or simple about God.

Life, in its highest sense, is humanity

Life, in its highest sense, is humanity, the very measure for all that is grand, is man.

Life reflects the Universe that is efficient

Life is not practical. Life is beauty, poetry, literature, art, music, drama, song, happiness, joy, love, honesty, care, science, creativity, productivity, honour, truthfulness, and many more that together have become the measure of the absolute of life, defined by man far beyond the practical, whereby life becomes efficient.

Life reflects the Universe that is efficient. Love is its quality that cannot be weighed, compressed, or quantified, but which flows in life as plasma flows in the Universe and powers every star.

Andromeda Galaxy

Galaxy NGC 5866

** Efficient Economics of the Universe
Consider economics. The Universe operates the most-efficient platform of economics that one can imagine. In the solar dynamic process, plasma is used up in vast quantities by every sun. Plasma is packaged into atoms that construct the planets. But in the very process of using up plasma, a process is set in motion that draws onto the scene more plasma than is being used up. By this self-expanding process, the galactic system expands as a whole. The Universe itself expands electro dynamically. The force of gravity plays no role in the larger context. It is a local force with local effects.

On the enormous scale

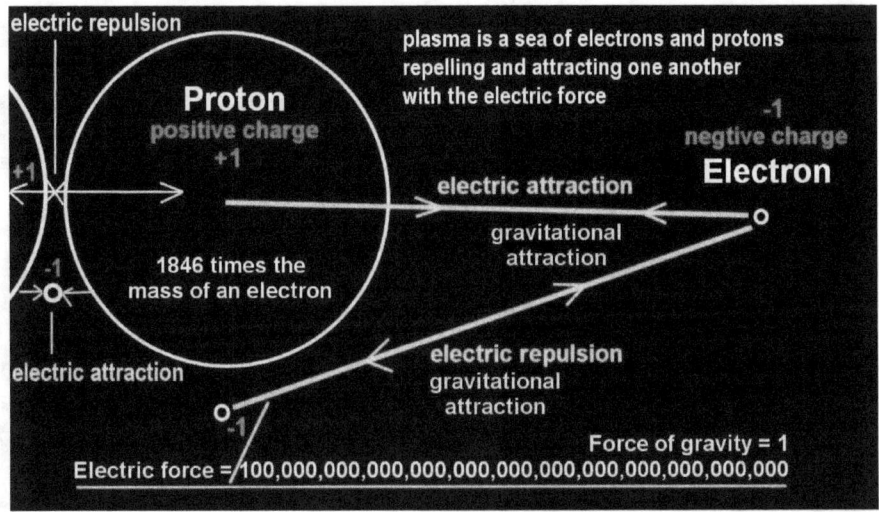

On the enormous scale, 99.999% of the mass of the Universe exists as plasma and is electric in nature, so that it interacts with the electric force that is 39 orders of magnitude stronger than the force of gravity.

 Atomic mass is extremely rare in the Universe. Only the planets and asteroids of solar systems are atomic in nature, and the fusion products of their respective sun. The remaining 99.999% of the universe is plasma. Streams of plasma in the universe are moved by the electric force, by which they extend across near infinite distances.

The force of gravity, in contrast is a field-force that diminishes with the square of the distance, because at double the distance, a gravitational field acts on a 4 times larger area. The force of gravity diminishes thereby so rapidly that it plays no role at all beyond the stellar cradle on, the immensely vast galactic scene.

Plasma streams are streams of discrete entities

The faucet
as a sink

Plasma streams in contrast, are streams of discrete entities that push each other onward by their electric interaction with one another. A plasma stream is comparable to a water pipe. A water pipe is filled with discrete atomic water molecules. It may have its source in the mountains. When a bucket of water is drawn from the pipe, all the water molecules in the entire pipe move towards where the water is drawn.

In space, plasma is drawn from the cosmic plasma environment in large streams by every sun. The sun consumes plasma in atom-synthesizing nuclear fusion. The consumption of plasma creates a sink effect that sets the entire plasma stream into motion.

In the galactic system the motion of plasma

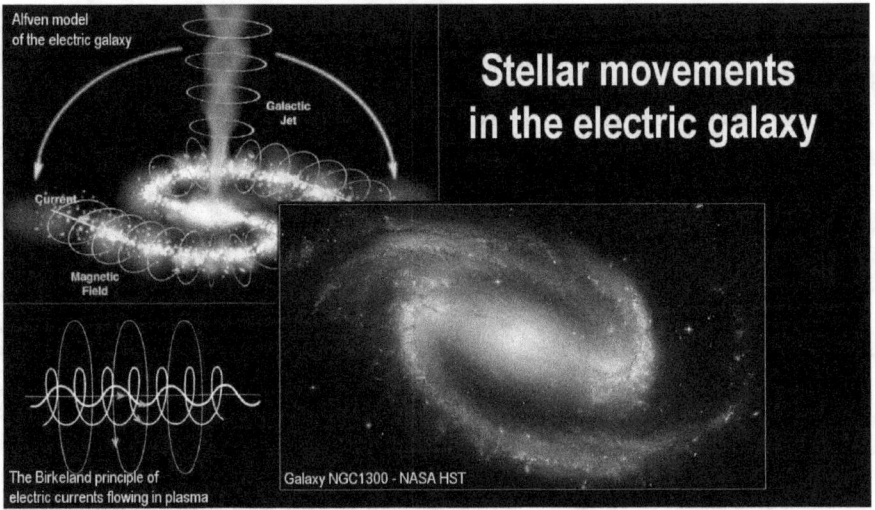

In the galactic system the motion of plasma is so massive that the resulting magnetic fields cause the entire galactic disk to rotate, slowly as this may be, and to cause the stars within the spiral arms to rotate axially within.

The Universe's Music Tone

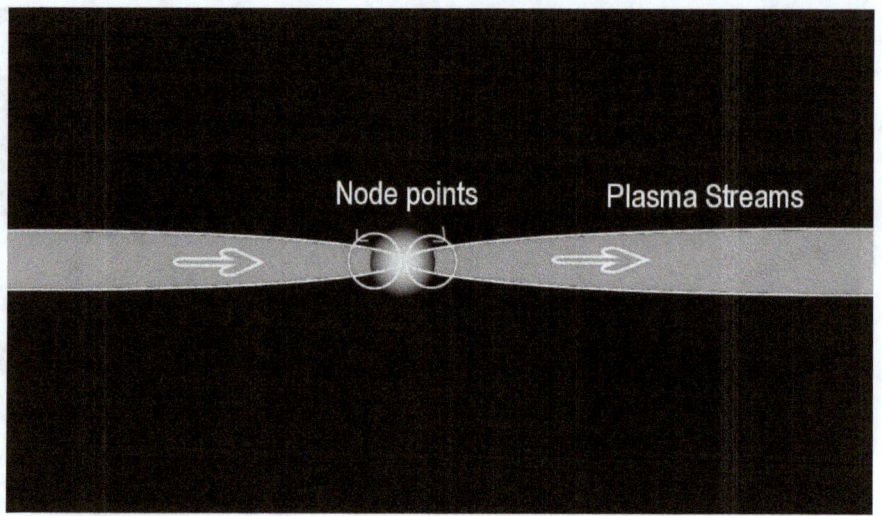

** The Universe's Music Tone

Also, as I said earlier, the movement of the electric plasma pinches the plasma streams themselves into discrete entities, and so, as discrete entities of mass with an electric charge, they begin to resonate in the wind that motivates them. They ring with a distinct tone, as it were, that is determined by their size and their density. The resonance, of course, is felt strongly at the node points where the effect is concentrated.

The principle of resonance

The principle of resonance is utilized in musical instruments. Different tones are produced by different length of strings, as in the case of the piano. A violinist produces different tones by pinching the strings against a plate at different distances from the sound bridge. In both cases, when the A note is played, the agitated string vibrates at a resonance frequency of 432 cycles of movement per second.

Plasma streams have their unique resonance 'tone'

Galaxies at node points

In a similar manner, the intergalactic plasma streams that connect galaxies with one-another across intergalactic space, have their unique resonance 'tone' that reflect the length of the plasma entity.

For the extremely long plasma streams

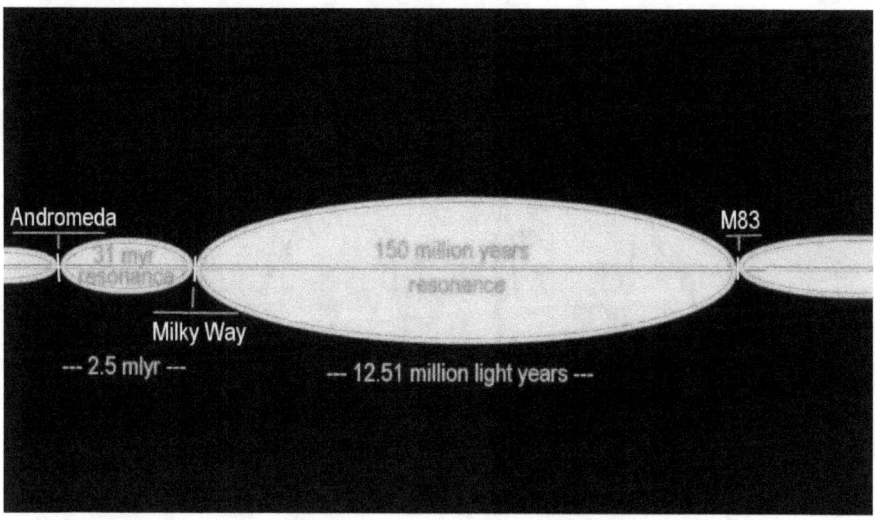

For the extremely long plasma streams that are extending across several millions of light years, their internal resonance tone is measured correspondingly in long time frames, typically in millions of years for a single cycle of 'vibration.'

For the case of our own galaxy, the Milky Way, one of the plasma streams rings with a frequency of 150 million years per cycle. The connection with the M83 Galaxy is hypothetical to illustrate a principle. The shorter connecting stream, hypothetically connects with the Andromeda Galaxy, resonates at 31 million years per cycle. The interacting resonance of the two very long cycles, affects the plasma density at the node point where the two streams come together, where our galaxy is located. The resulting plasma density fluctuation affect the entire galaxy over these long time frames, which, of course, also affects the climate on Earth with climate fluctuations occurring over the same long time frames.

Two long-term climate cycles have been found

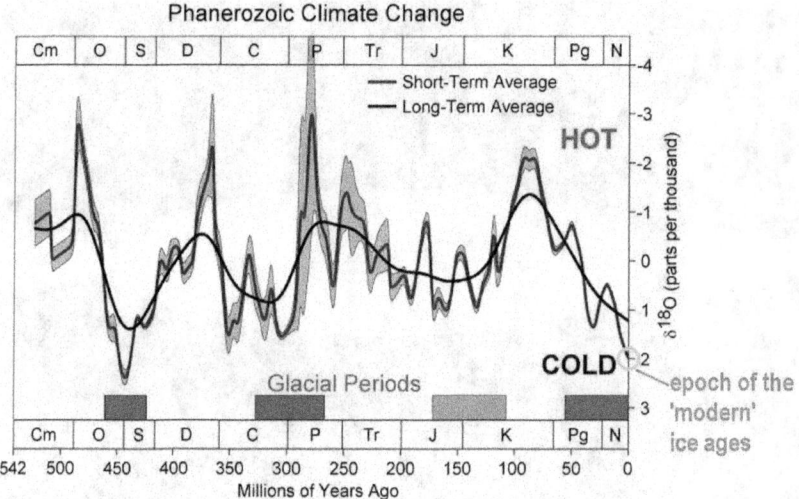

Phanerozoic Climate Change

Evidence for the two long-term climate cycles have been found in ocean sediments. There, we see both tones of the intergalactic plasma streams coming together, being overlaid on one-another. The combination of these tones, in more recent time, has caused the climate to cool so extensively that Antarctica froze up roughly 35 million years ago and remained frozen for few million years until the upswing of the short cycle caused it to thaw again.

Both, the long and short cycles, are approaching their low point

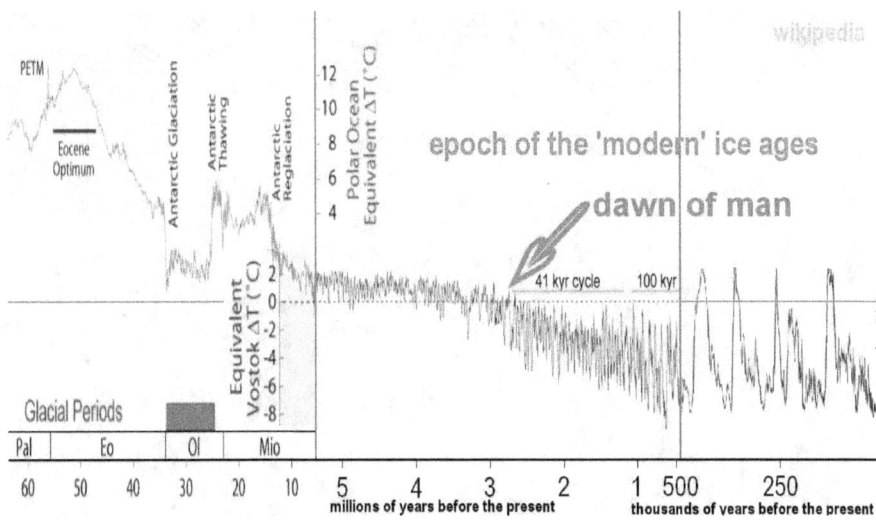

With the next down-swing of the short cycle Antarctica froze up again 12 million years ago and remain frozen. We are now at the stage where both, the long and short cycles, are approaching their low point. In the course of the weakening plasma density in the galaxy, with it being affected from both sides, the epoch of the ice ages began roughly 2 million years ago, which will likely continue for a few more million years till the next upswing of the 31-million-year cycle begins an takes effect.

Our electric universe is a dynamically operating complex of great power. Naturally, the interacting dynamics also produce correspondingly great effects. In addition, as one would expect, the principle of the dynamics that are expressed on the galactic scale, are also expressed on the small scale of the solar system which has its own plasma resonance effects that further affect the climate on the Earth. The ice age cycles result from that.

The interstellar resonance tones

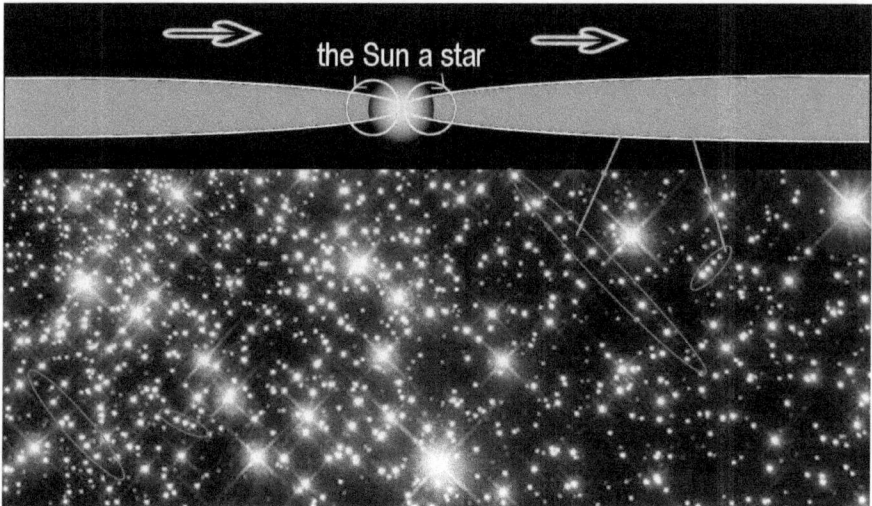

Because the stars within the galaxy are similarly interconnected with each other with interstellar plasma streams, the long galactic resonance tones are further modulated with the interstellar resonance tones that have a frequency of one cycle for every 120,000 years. The numerous glaciation periods of the ice ages in the last two million years result from the effect that the interstellar resonance has on our Sun. As the result of this interaction, the glaciation condition produced by the long galactic cycles, get periodically interrupted with the nicely warm interglacial periods, as the one that we presently enjoy, which, unfortunately is ending.

What does this all mean?

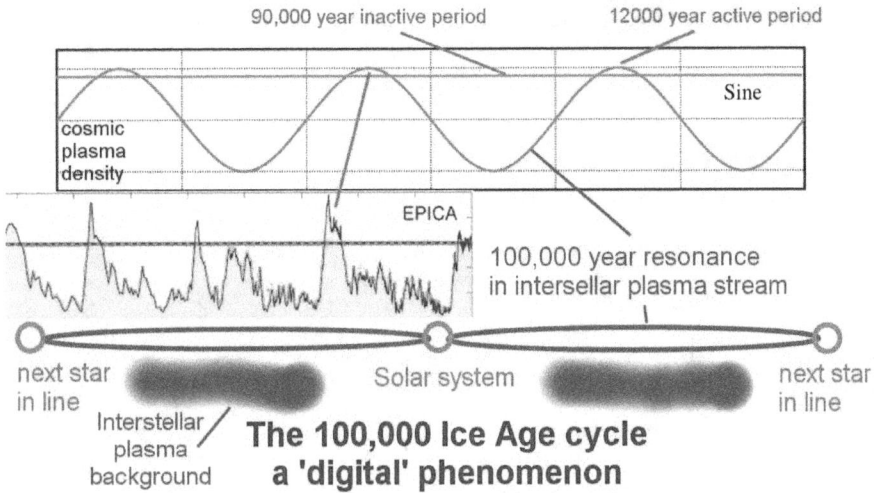

The 100,000 Ice Age cycle
a 'digital' phenomenon

What does this all mean? It means for us that with the galactic plasma density being at a historic low point, the interstellar resonance that draws its plasma from the galactic environment, drops for long periods below the minimal threshold line that determines the plasma density that is required for the Sun to remain active. When the plasma density diminishes below the threshold line, an Ice Age glaciation period begins. The glaciation period typically lasts for 90,000 years. All evidence tells us that the Earth is presently in the boundary zone towards the Sun going inactive and for the next Ice Age to begin, which is likely to occur in roughly 30 years going into the 2050s.

The process that is unfolding here is similar to what we see happening in economics that is in a transition towards total collapse that happens when the threshold line is crossed. In the solar system the plasma weakening is only minutely felt. A system of internal regulation that is reflected in the solar-wind pressure, keeps the solar fusion process operating steadily to within a fraction of a percent of its output level. But when the reserves for this built-in

111

regulating system become exhausted, the entire process disintegrates in a chain-reaction collapse. At this point the dynamic primer fields collapse that focus plasma onto the Sun.

When a threshold is crossed in recognized economic value

The equivalent stage in economics can have many causes. When a certain threshold is crossed in recognized economic value, nothing works anymore. The interlinked processes no longer function. For example, when the value of money becomes so uncertain that the price for products become meaningless, as no one knows what the value of money is or will be in an hour, then the stores simply close, the banks close, the electricity supply stops, the gas pumps stop, food is no longer delivered, cooking food stops, people starve and die, or shoot each other for food. The collapse chain-reaction may begin one morning when traders in financial markets can no longer trust the value of money, because, when nothing much remains standing in economic products that the value of money inherently represents, then the value system is dead. At this point, when the bare reality becomes obvious, the entire commercial process simply shuts down. This may take only hours, and by nightfall the system is dead. This potential exists.

The solar system can collapse just as quickly

Ice Age of the dimming Sun in 30 years

www.ice-age-ahead-iaa.ca

The operation of the solar system can collapse just as quickly. When the primer fields collapse that focus plasma onto the Son, the interglacial period is over. The difference is that unlike the solar collapse into an Ice Age, which is cosmic in nature and cannot be avoided, the economic collapse process is manmade, resulting from insanity in policies, so that the collapse process can be reversed with saner policies. But will we do this? Also, will we make preparations for the coming Ice Age? Our existence hangs in the balance on both counts.

The astrophysical 30-year estimate

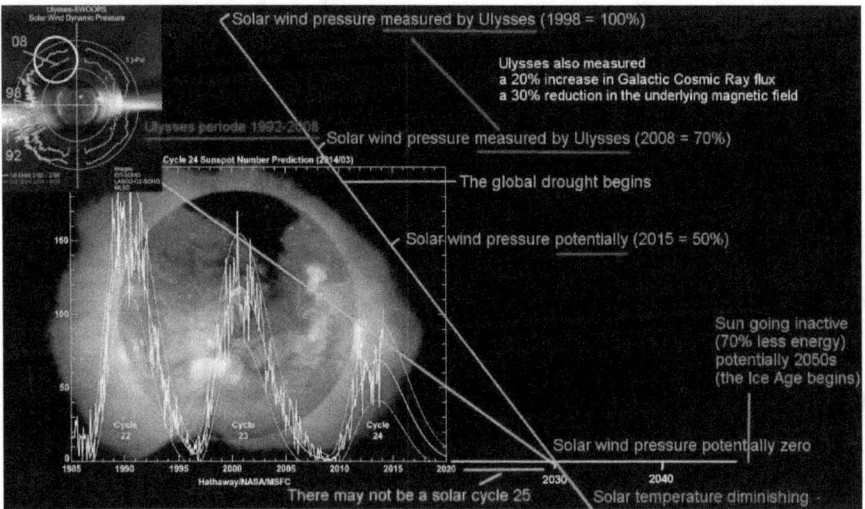

The astrophysical 30-year estimate reflects the enormous weakening of the plasma density in the solar system that has been measured in numerous ways, from space with measurements of the solar wind pressure, on the ground with simple solar cycles observation, and in the Arctic with magnetic-pole drift measurements, and so on.

Our response to the measured weakening dynamics, in conjunction with the measured consequences in historic ice core samples that suggest a 70% reduction in radiated energy from the Sun, should be, to shift our living into the tropics before the enormous Ice Age transition begins that renders most areas outside the tropics uninhabitable in an extremely short time once the Sun goes inactive.

*The factor of practicality

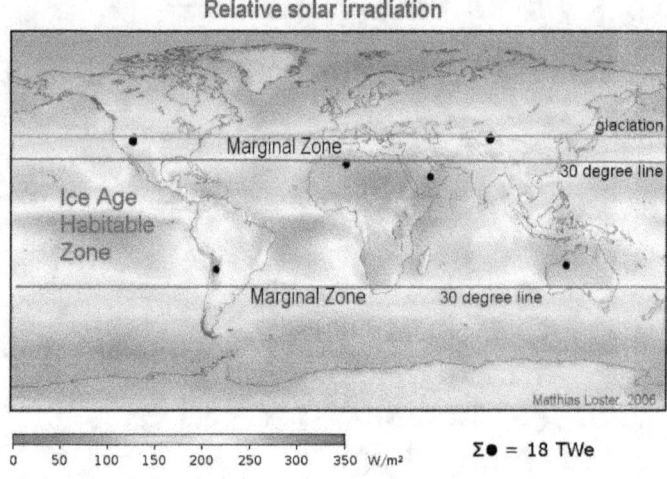

Here the factor of practicality comes to the foreground again. While we lack the means to affect the vast cosmic processes that act on the climate on Earth, we do have the power to affect our actions as a society in response to cosmic dynamics that we have scientific evidence of, even to the point of understanding the consequences. But will we act?

Relocate Canada, Europe, Russia

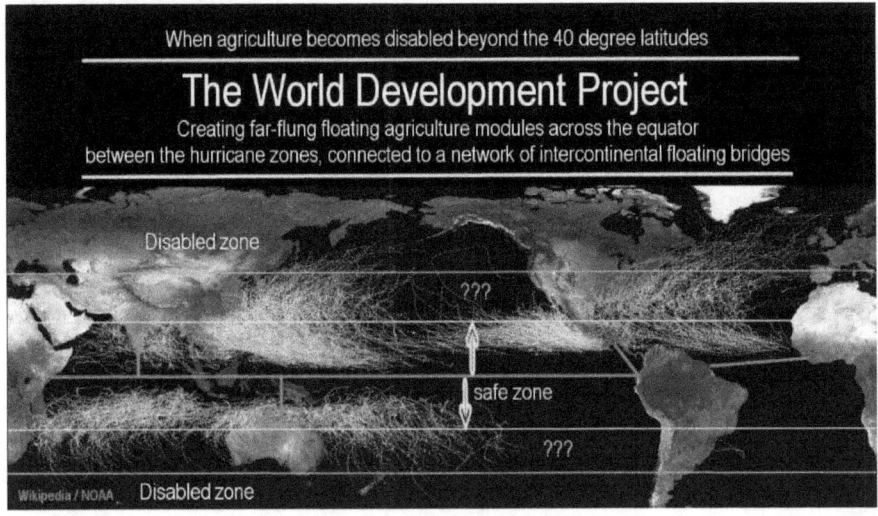

It is obviously considered totally impractical in the present small-minded world, to even consider to relocate Canada, Europe, Russia, and so forth, into the tropics, with nothing more practical than scientific data and scientific recognition as an imperative. This means that today's small-minded practicality that has become a near-global phenomenon, will most likely cause humanity to become extinct when the world's food supply infrastructure that is keyed to the warm interglacial climate, becomes disabled worldwide without new infrastructures having been created that enables the continued existence of humanity.

The design of the human species and its continuing development is as big in power and potential as are all the systems of the Universe. We have gigantic capabilities at our finger tips. We have the scientific and technological capability to shift the entire global food supply onto indoor agriculture, as an option. Likewise, we have the technologies and resources at hand to create a system of floating agriculture spread across the equatorial sea, extending from world-encircling equatorial sea bridges. Building such infrastructures would be an efficient response to the now ongoing Ice Age transition. It would be efficient, because it would enable humanity to continue to live, which would otherwise become likely extinct. That's an extreme example of the difference between efficiency and practicality. On which side will we weigh in?

The 'practical' approach to humanity's future is presently determined by a transnational club of liars, swindlers, and thieves who project the myth of manmade global climate change and are fast destroying under this myth the civilization that humanity has built. Their lies are now shaping the human response. The lies keep the vision of society small and narrow, while they subject the world to looting, poverty, starvation, and war under their club's long-standing depopulation doctrine that is deployed to protect the club's empire that is essentially already dead. This is the sad face of the practical world. On this platform of extreme 'practicality' the world's zero-response to the Ice Age imperative is almost guaranteed. Whoever raises to topic is considered an idiot. Countless excuses are being dredged up to make the Universe and the human being appear small, docile, and simplistically unchanging. Consequently, nothing is being built. Nothing is even considered. Just ask anyone.

The focus in practicality

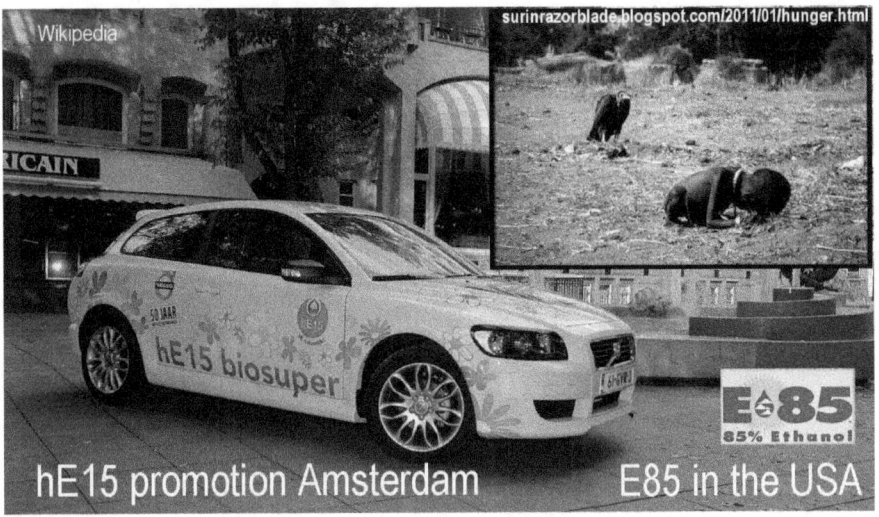

hE15 promotion Amsterdam · E85 in the USA

Debt, profit, and war, are the issues of the day, with the depopulation policy standing in the background for which society has been educated to massively burn its own food resources in automobiles as biofuels - a process by which 100 million people are starved to death each year, with nearly everyone taking part in the murdering in some form.

The focus in practicality is not on protecting humanity, but is on destroying it. The cosmic Ice Age, and the science for its timing, are no longer issues being considered in the dying landscape of small-minded dreaming.

We, as human beings, as a part of the Universe

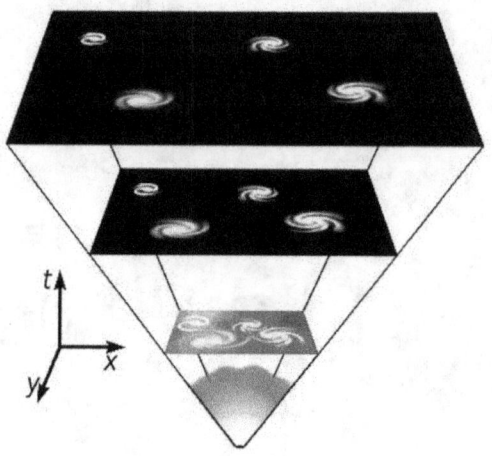

It might be possible that our exit from this small-minded trap of dreaming can be inspired by our growing scientific recognition of the dynamics of the Universe that are immensely efficient, and unimaginably powerful and creative, which contrary doctrines are designed to hide. While the club of the small-minded of the Old World would have us believe that the entire universe exists as atomic matter that was born in a hypothetical Big Bang explosion, for which no evidence exists, from which the universe is deemed to have condensed into galaxies of stars and clusters of galaxies that are powered by their own energy within, which is now winding down, the real evidence that does exist points to far-more efficient realities. It points to a universe that is dynamically its own inexhaustible energy, by which it is building itself, and is progressively expanding, quantitatively, and is developing qualitatively. We, as human beings, as a part of the Universe, are our own evidence of the process with our scientific understanding that takes us efficiently far beyond the visible and 'practical' on a path of efficient development that enables us to see the Universe evermore richly.

A Universe that doesn't need the Big Bang

The center of the Milky Way, at the center of the Big Bang explosion of the universe

We see a Universe that doesn't need the Big Bang of the small-minded origin theory. We see a Universe that doesn't merely produce the energy by which it exists, but is a Universe that IS energy itself.

Theoretical physicist, David Bohm

The leading-edge theoretical physicist, David Bohm, whom Albert Einstein had recognized as his successor, describes a universe that is not primarily empty space as we seem to experience by flying spacecraft in this space, but is itself a vast sea of inherent energy that has an efficient implicate order and expressed explicate order. The evidence for David Bohm's concept of an implicate order may be found in the specific propagation-speed of light that is not inherent in the light itself, but is the same for all 'colors' and for all levels of energy from which the light is projected.

Ripples of energy

Ripples on water

www.mrwallpaper.com

The evidence for what David Bohm referred to as an explicate order, which he likened to ripples of energy similar to ripples on the surface of the sea, may be the source for all the electrons and protons in the Universe that flow together into the plasma streams that power the galaxies and all the stars within them that create their planets with the fusion of plasma into atomic elements.

Electrons and protons are not basic entities

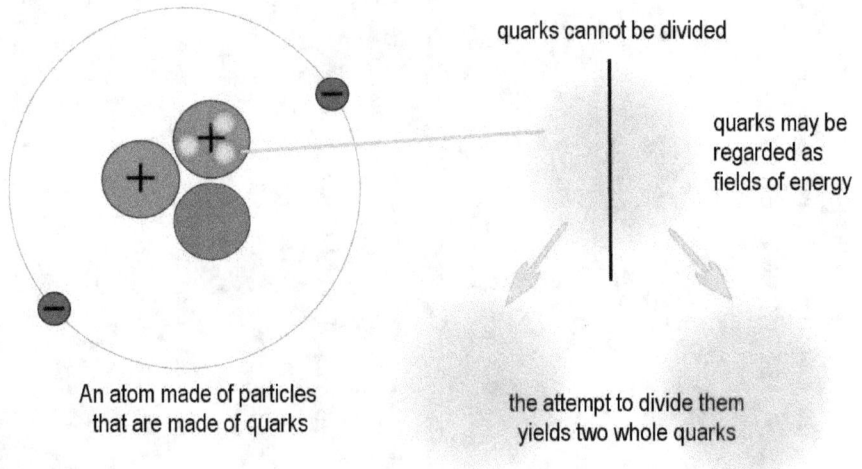

An atom made of particles that are made of quarks

quarks cannot be divided

quarks may be regarded as fields of energy

the attempt to divide them yields two whole quarks

It is well understood in nuclear physics that the electrons and protons are not basic entities themselves, but are themselves the constructs of combinations of quarks that are recognized in nuclear physics as organized moving points of energy.

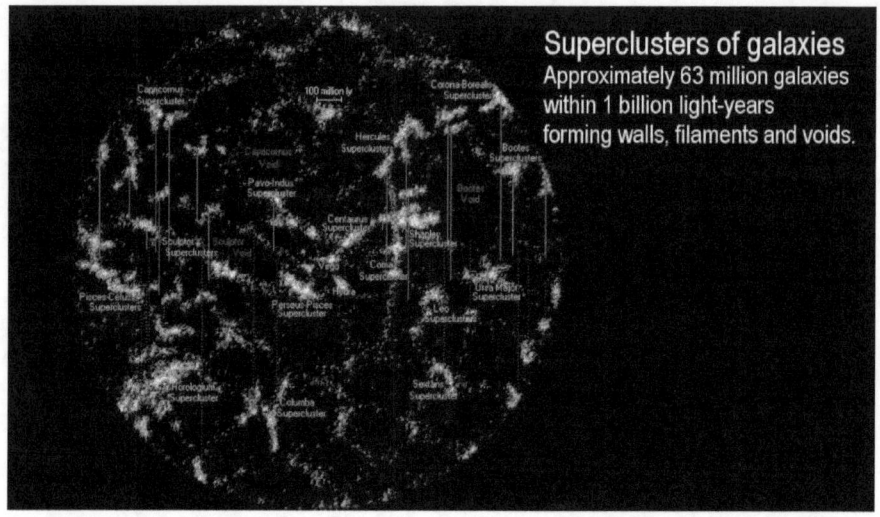

Thus, the recognition that we live in an extremely efficient Universe has, as a concept, an amazingly solid foundation.

On a planet under a rather mediocre sun

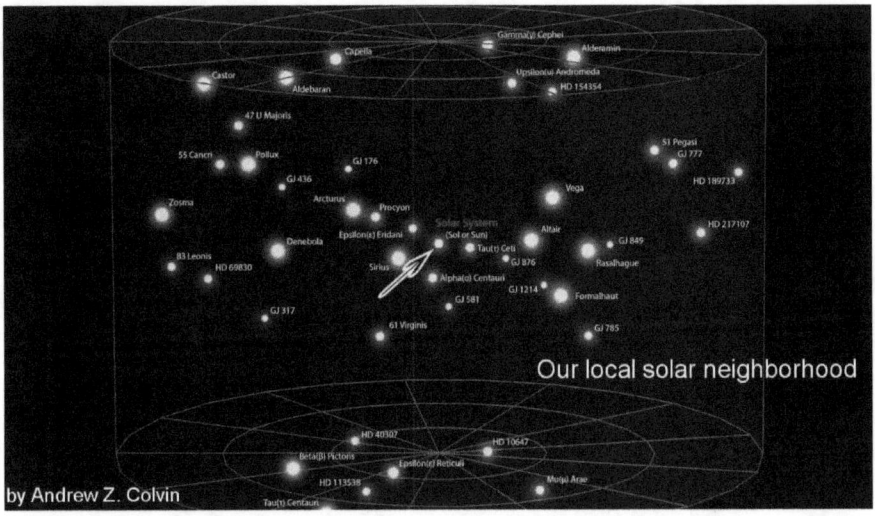

Our local solar neighborhood

by Andrew Z. Colvin

Of course, what unfolds dynamically from this foundation, has some major practical implications for our living on a planet under an electric Sun that is a rather mediocre sun, which in edition is located in a thinly populated area of our galaxy. Both factors render our Sun susceptible to marginal conditions, as we presently experience in the form of Ice Age epochs and cycles, and interglacial periods. It could well be, in this context that the local interstellar plasma streams, which do carry a large physical mass and have a built-in resonance that has a major effect on our Sun, may be the reason why we exist at all.

Interstellar resonance raises the galactic background

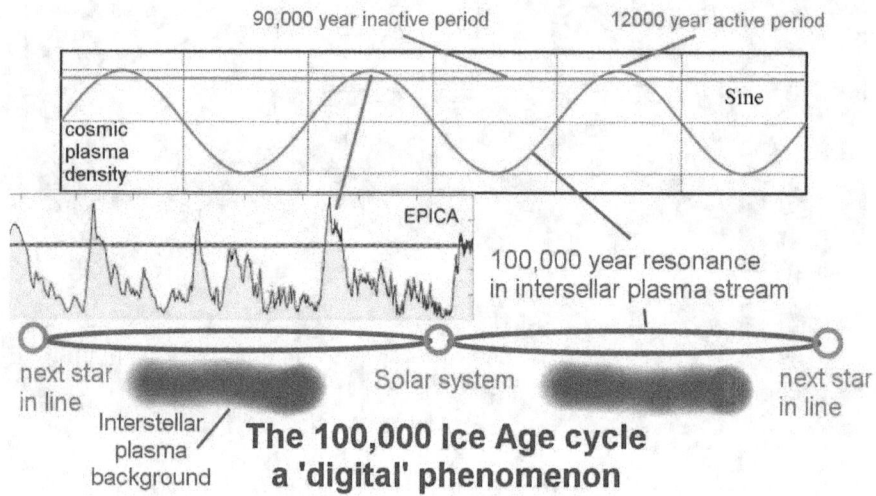

The 100,000 Ice Age cycle
a 'digital' phenomenon

The interstellar resonance has the effect that it raises the galactic background plasma density above the solar cut-off threshold periodically, which interrupts the galactic ice epoch and enables the amazingly warm interglacial holidays, like the one that we presently enjoy, which stands as an anomaly in the dark glacial environment with an inactive Sun that has been the norm for 85% of the last two million years.

This means that the ice ages that render the conditions for life on our planet so harsh that only a few million people have emerged from the last Ice Age, are not the result of interstellar oscillations, but are the result of intergalactic oscillations, while the interstellar oscillations effectively boost the solar system periodically above the galactic background, which gives us a holiday on Earth to recover ourselves.

It may well be, that without these recovery periods from the long glacial night, we might not have had the means developed to exist. As it was, in spite of it all, quite a few distinct human species have become extinct in the long sweep of human development that occurred entirely during the galactic glacial epoch.

We are the eighth human species

Main human species

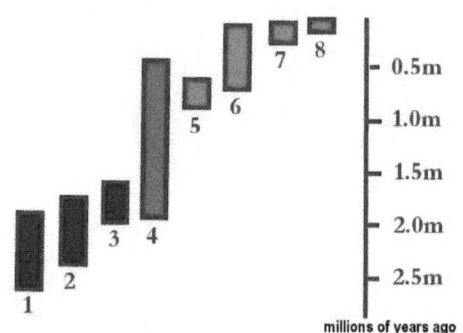

australopithecus rudolfensis (1),
australopithecus habilus (2),
homo ergaster (3),
homo erectus (4),
homo antecessor (5),
homo heidelbergensis (6),
homo neandertalensis (7).
homo sapiens (8)

0.5m
1.0m
1.5m
2.0m
2.5m

millions of years ago

We, the homo sapiens (8), are the only surviving,
and the shortest lived of all the the human species,
at barely 200,000 years of age.

We are the eighth human species, and the only remaining one with a mere 200,000-year history. The other species may have all perished in the unfolding long glaciation cycles, even while some, like homo erectus, had survived a very long time. The high rate of extinction may be the result of the potentially fast transition to ice age conditions when the Sun goes inactive. Without being prepared for such results the surprised societies would likely have been decimated to extinction by the changing conditions. We may suffer the same fate in out time if we allow this fate to come upon us by keeping our eyes closed and our mental vision confined by a shallow sense of practicality.

But why should we fall into this trap when the evidence that we live in a plasma- electric Universe, is all around us?

** The Ecliptic

That the universe is completely and fully electric in nature at all levels, is evident by another major feature of evidence of electric dynamics that is found everywhere, from the galactic system, to the solar system, and is even reflected in planetary systems. A typical galaxy, for example, exists in the form of a thin flat disk of stars, often with a slight bulge at the center.

For our Milky Way Galaxy

A Milky Way look-alike NGC 6744

"Wide Field Imager view of a Milky Way look-alike NGC 6744" by ESO - under CC BY 3.0 via Wikimedia Commons

For our Milky Way Galaxy, for example, the disk of stars is estimated to be between 100,000 and 180,000 light years wide, containing a field of approximately 100 billion to 400 billion stars, which are all solar systems of various types.

A disk-like planetary ring system

Saturn eclipsing the Sun

We've seen this when we looked at Saturn, for example. Saturn is surrounded by a disk-like planetary ring system. The disk is a mere 20 meters thick, but is 120 million meters wide.

As is plainly apparent

Also, as is plainly apparent, Saturn's thin ring system is perfectly aligned with the planet's equator.

Also the ecliptic for Saturn's moons

The ring system is also the ecliptic for Saturn's moons that all orbit on the same plane or are huddled close by.

Do we see the same force expressed here, that on the larger scale compresses a galaxy into a thin, flat disk? The similarity suggests that this is indeed the case.

A highly simplistic view

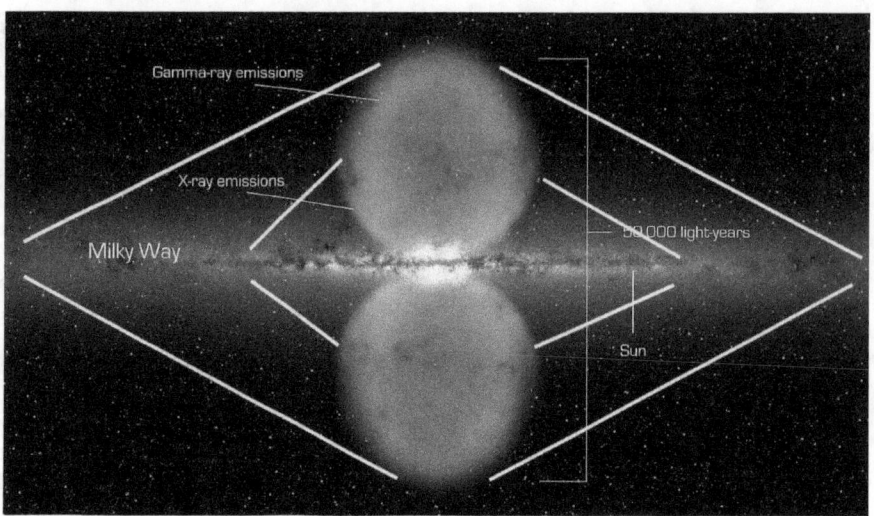

The concept presented here is a highly simplistic view of electromagnetic compression from two opposing directions, to illustrate a principle.

Ecliptic effect is likely magnetically produced

David LaPoint: The Primer Fields   Experimental, by A. Peratt

The real ecliptic effect is likely magnetically produced by the very large galactic primer fields. In an experiment, the researcher David LaPoint, placed a number of non-magnetized steel balls on a sheet of glass between two bowl-shaped permanent magnets of opposite polarities, similar to what one would encounter in cosmic primer fields. In the experiment the steel balls spaced themselves magnetically apart from each other and formed two concentric circles.

In another experiment

David LaPoint: The Primer Fields

In another experiment David LaPoint placed two of the steel balls on a sheet placed in line with the axis of the two magnets. The result was that the two balls, no matter where they were placed would seek out the line of the ecliptic and would oscillate along this line until they reached a specific point.

Another set of evidence

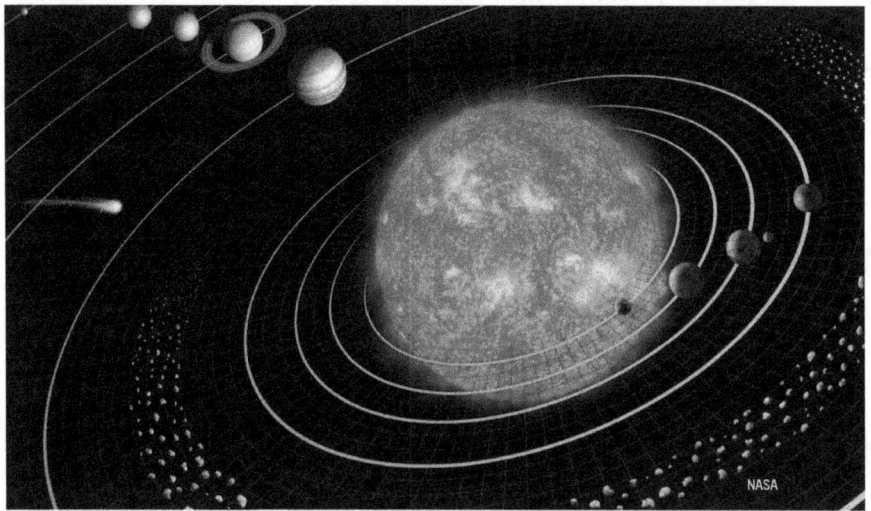

While David LaPoint's experiments do not prove that the galactic, solar, and planetary ecliptics are electromagnetically organized, the experiments nevertheless suggest such a potential, especially if one considers that these static experiments agree with observations made in high-power dynamic experiments, as indicated in the previous slides.

There also exists yet still another set of evidence that suggests that the solar system's ecliptic plane, and the ordering of the planetary orbits is electrically organized and has been electrically maintained over extremely long periods. The potential proof lies in what would happen if the ordering effect of the primer fields would be removed, as would be the case during the glaciation periods when the Sun is inactive. What would one expect to happen? The planets would maintain their orbits by their mass, gravity, and orbital velocity interplay. All asteroids would do so likewise. However, with the small asteroids having a large drag to mass relationship, their velocity would dwindle over time, and so would the radius of their orbit. As a result, one would expect to see radically increased asteroid impacts on Earth towards the end of each glaciation period. Not surprisingly, this is exactly what we see in the ice core samples from previous glaciation periods.

In Antarctic ice samples from the four last ice ages

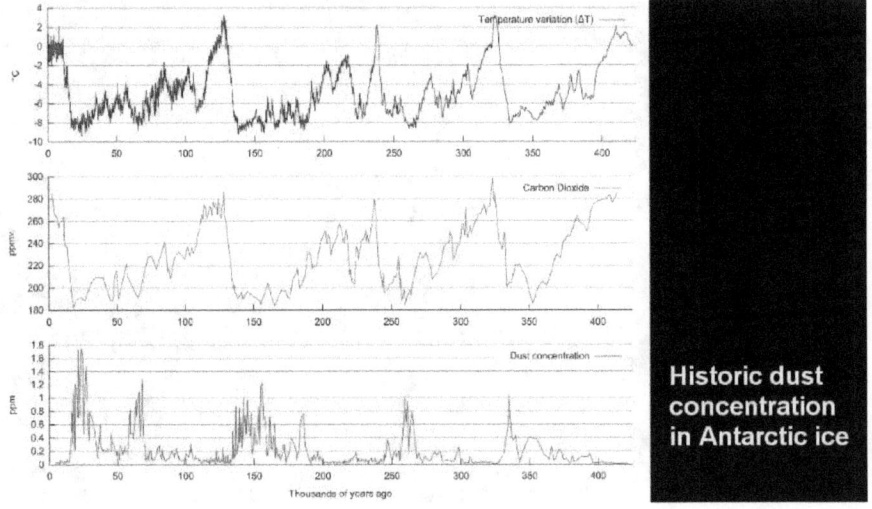

Historic dust concentration in Antarctic ice

We see in Antarctic ice samples from the four last ice ages, large dust accumulations occurring in the latter part of each of the glaciation periods. We also see that as soon as the interglacial begins, the dustiness suddenly stops, in every case. While this doesn't prove that the phenomena were caused by missing primer fields during the glaciation periods, the potential that this might be so is nevertheless interesting.

141

Globular clusters

Globular Cluster M10
(ESA/Hubble)

Another potential evidence for the ecliptics being caused by the primer fields, is found in areas of the galaxy where large congregations of stars exist outside the normal dynamics where large-scale primer fields should not be found. In these areas stars have congregated in globular clusters. The clusters are spherical in nature, as one would expect for non-organized congregations of stars.

Globular clusters appear to be congregation of stars

Globular cluster Messier 54 by ESO - CC BY 4.0 via Wikimedia Commons

Globular cluster NGC 104 by the Southern African Large Telescope (SALT) CC BY 3.0 via Wikimedia Commons

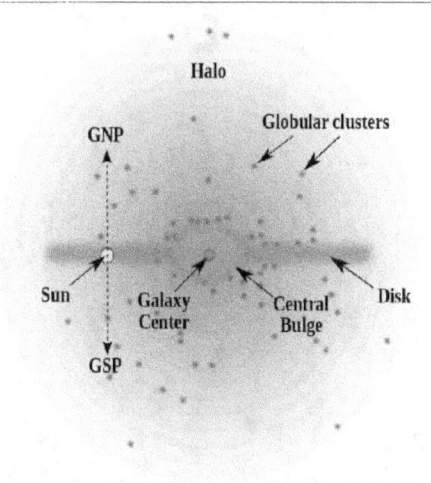

"Milky way profile" by RJHall - CC BY-SA 3.0 via Wikipedia

Globular clusters appear to be congregation of stars that were swept out of the spiral arms by the vast tangle of plasma currents that surround the galaxy like a halo. Observations indicate that all galaxies have them. For our Milky Way Galaxy, 150 such clusters have been identified, with up to a million stars in some, as that of the lower image. The larger Andromeda Galaxy may have 500 star clusters, and some very large galaxies at the center of large clusters of galaxies are believed to have as many as 13,000. But they all have one thing in common. Their congregations of stars are spherical. They are not compressed into ecliptic structures. This is precisely what one would expect for these fringe features that operate outside the big galactic plasma streams.

Of course, the coincidence of the lacking ecliptic with the lacking primer fields cannot be cited as proof that we live in an electric, plasma defined, universe. But it proves something in conjunction with all the other types of evidence that point to large-scale electric phenomena, such as nebula that are obviously electric phenomena.

Gas clouds surrounding hyper-active stars

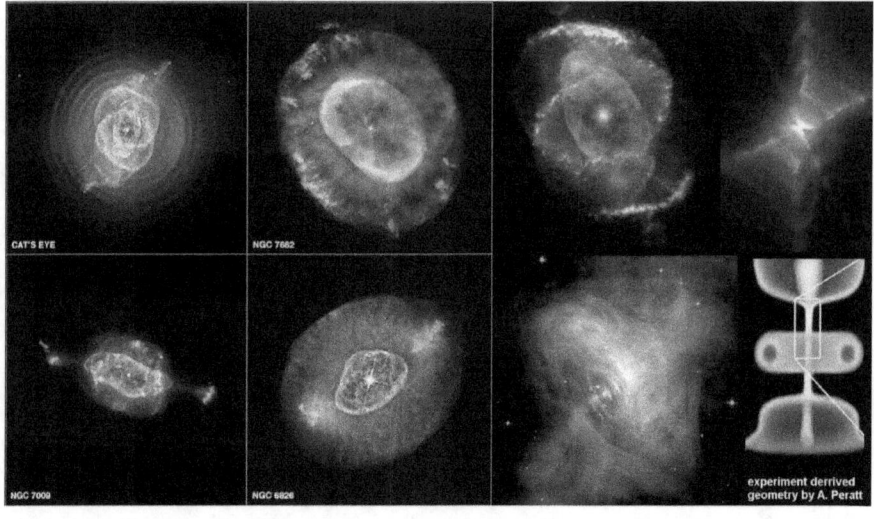

These immensely luminous phenomena are evidently not hotly glowing gas clouds energized by star explosions in distant times that have continued to glow unabated. If this was the case their energies would have dissipated eons ago. These gas clouds surrounding hyper-active stars glow brightly for the simple reason that they are continuously powered by large interstellar plasma streams. Their bipolar feature leaves little doubt about their nature. Some day, no doubt, they will be recognized as that. At this point astronomers will no longer insist that plasma does not exist in cosmic space, and proclaim that all phenomena in the Universe must be the effect of gravity, as is the current belief. When the belief is shed, astronomers will be shocked to realize that they had been missing the very point that is foundational to every facet of astronomy and is absolutely crucial for the future of human existence on the Earth. The breakout point from the small sphere of practicality, to the gigantic sphere of universal efficiency, may not be far distant, because in the realm of the mind no inertia exists that should hold back the movements of discovery that may have just barely begun with a wide open horizon before us.

The evidence is so wide and extensive that we live in an electric universe, expressed in the galactic system, the solar system, and the planetary system, that one actually has to make a major effort to close one's eyes and mind to the reality that is increasingly hard to deny.

I would like to suggest that we are not the same people anymore who came out of the last Ice Age with a minuscule world population and a primitive sense of reality, but find ourselves evermore with open eyes and an open mind, and with science and technologies that can change and enrich the world as nothing ever had.

I would like to suggest that we have developed ourselves into giants, comparatively speaking, during the last 10,000 years of the current interglacial recovery period. Only during the last two centuries did we become infantile and practical again, and began to babble in fairy tales.

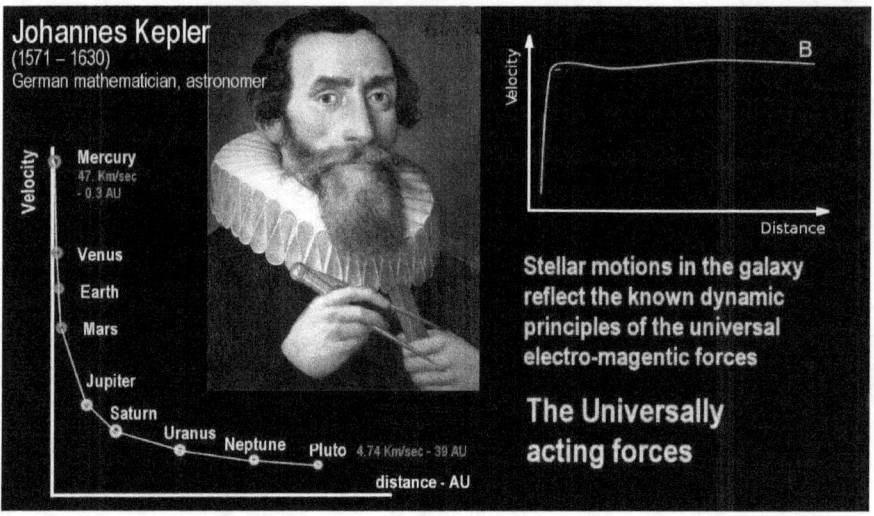

Johannes Kepler
(1571 – 1630)
German mathematician, astronomer

Velocity

Mercury
47. Km/sec
- 0.3 AU

Venus
Earth
Mars
Jupiter
Saturn
Uranus
Neptune
Pluto 4.74 Km/sec - 39 AU

distance - AU

Velocity

B

Distance

Stellar motions in the galaxy
reflect the known dynamic
principles of the universal
electro-magentic forces

The Universally
acting forces

The great astronomer, Johannes Kepler of the early 1600s, who put science on the map as a rigorous discipline, would cry in his grave if he could hear us today spinning the yarn of such absurd notions as orbiting stars, and of the density-wave theory that is dreamed up as an epicycle to make the absurdity seem plausible. Kepler would cry, "have you learned nothing from me?" Why do you violate my name so brutally as to suggest that the stars in the galactic plane are orbiting around the galactic center, which is physically impossible by the laws of orbital motions that I have spent a lifetime to discover for you? The relationship of orbital distances versus orbital-velocity, which is inherent in all orbital motions, is not physically possible on the galactic scale even with all the epicycles added that astronomers like to apply. Thus, the actually measured evidence in stellar motions disproves the orbital theory."

Kepler would ask further

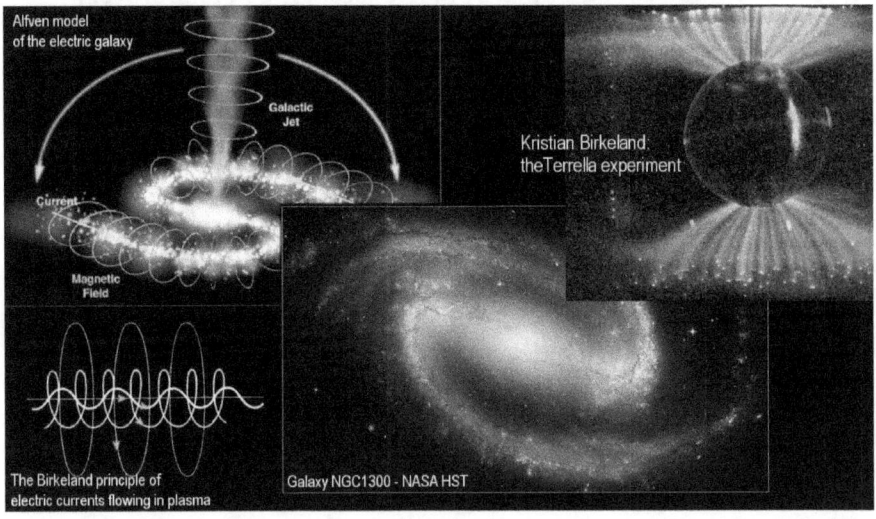

Alfven model of the electric galaxy

Galactic Jet

Kristian Birkeland: the Terrella experiment

Current

Magnetic Field

The Birkeland principle of electric currents flowing in plasma

Galaxy NGC1300 - NASA HST

Kepler would ask further, "why do you also spit in the face of your own modern pioneers, like the Nobel laureate Hannes Alfven who gave you the answer in principle, that you seek, that would clear my name and heal your folly? Alfven proposed the only possible type of answer to the movement of stars in a galaxy, as expressions of the magnetically induced rotary motions of free flowing electric currents along the length of the spiral arms. The self-aligned twisting currents were predicted in 1908 by the Norwegian explorer and physicist Kristian Birkeland. The existence of these currents was confirmed with satellite measurements in space in 1967. The nature of the currents was further replicated by Birkeland in his Terrella experiment, which illustrates that the Birkeland currents occur in groups of field-aligned current sheets. Modern measurements of the movement of stars actually confirm Alfven's perception of the Birkland rotation in the plasma environment in the spiral arms of galaxies. The terrestrial Sun has been measured to be moving upwards towards the edge of the galactic disc, instead of laterally in an orbit. The resulting plasma-rotation concept for the 'circular' stellar motions within a spiral arm, is efficient. It stands on a solid, truthful foundation and on proof in several types of laboratory experiments. Shouldn't science be open to what is efficient, truthful, and is built on discovered principles that can be demonstrated in laboratory experiments, and so forth?"

Without truthfulness in science

Annihilation is assured

500,000 times
Hiroshima
in one hour

Castle Bravo - the first U.S. test of a dry fuel thermonuclear hydrogen bomb - March 1, 1954 at Bikini Atoll, Marshall Islands

Indeed, without truthfulness in science, how can we clear the logjam in political theory that defines the potential extinction of humanity in nuclear war as a practical foundation for civilization?

Efficiency in science means

Photo by Scott Williams

The Supreme Being

wikipedia

I would like to suggest that we have developed ourselves not only technologically, but also as human beings, as we have developed our higher-level human resources towards ever-greater efficiency in scientific self-recognition. Efficiency in science means movement without inertia in thought. It means progress without limits in time. It means that we have the capacity to be were we want to be. It means that we will find ourselves with the power to vacate the threat of nuclear war from the fabric of civilization, and this with the full sweep of vacating all the destructive forces of empire that war is a part of, like the self-collapsing economies, the swindlery in finance, the green fascism in environmental concerns, and the looting of the world for depopulation, which together have staged a deadly destructive path on every front, and have even crippled the window of science in many respects for the same intention. Efficient living means vacating the very notion of empire, with which the freedom of humanity begins.

A new stage for the 30 years we have remaining

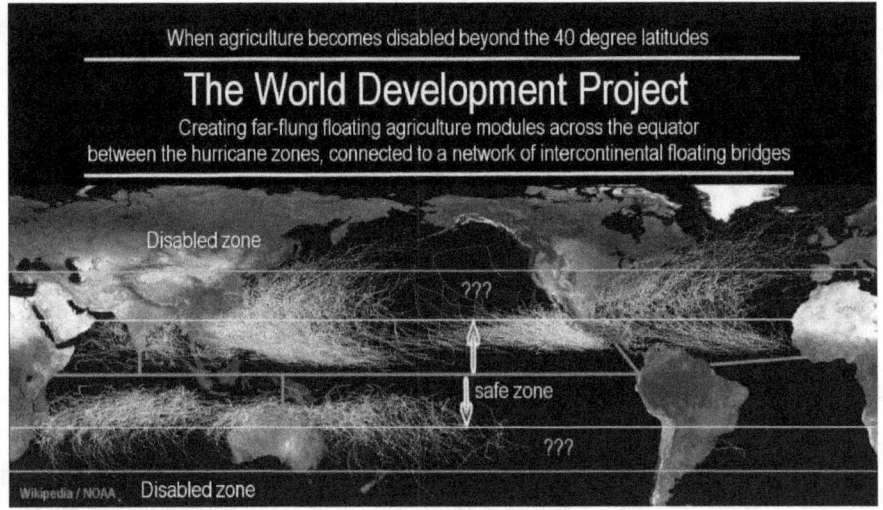

I would like to suggest that we will step away from the small-minded practical insanity and set a new stage for the 30 years that we have still remaining, potentially, until the next glaciation period begins, and that we will use the great human potential that we have at hand to cause a profound world development on a global scale with such efficiency that we assure ourselves thereby that we will live through the next Ice Age, and live richly and abundantly in spite of the harsher climate conditions that come with the Ice Age that we cannot avoid anyway.

Since the potential that we will do all this is greater than we like to believe. I would like to suggest that we stand not at the precipice of doom in our time, but stand at the portal to the greatest era of discovery and efficiency in living that we have yet imagined, so that when the next ice age knocks at the door, we will answer, where is your sting?

This is what it means to live as human beings

I would like to suggest that this is what it means to live as human beings, and not merely as a practical people, but as an efficient people on a scale far beyond anything that we yet dare to dream of.

www.ingramcontent.com/pod-product-compliance
Lightning Source LLC
Chambersburg PA
CBHW070247190526
45169CB00001B/330